PAPUA NEW GUINEA

Mathematics

Essential Skills

7

Pat Lilburn

Contents

For Students

The *Oxford Essential Skills Book* has been written to support the *Oxford Grade 7 Mathematics Student Books A* and *B*.

The focus of this book is on providing practice examples to revise and support the concepts learned in the two Student Books.

The book is set out under the five Strands outlined in the *Upper Primary Mathematics Syllabus* and *Teacher Guide*:

Number and Application
Space and Shape
Measurement
Chance and Data
Patterns and Algebra

Within each Strand, you will find many practice examples that relate directly to the Learning Outcomes for that Strand. These Learning Outcomes are specified in the *Upper Primary Mathematics Syllabus*.

The subheading in each Strand shows you what the examples will help you practise.

7.1.3 Convert between fractions, decimals and percentages

The Help Box contains worked examples to show you how to set out and complete the examples in each Strand. The Remember Box contains information that will help you complete the examples.

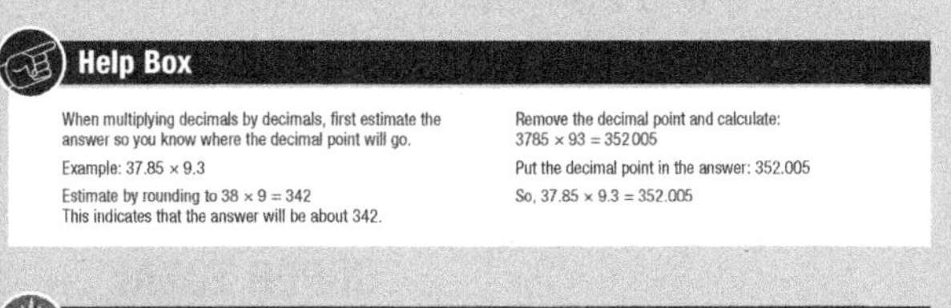
Help Box

When multiplying decimals by decimals, first estimate the answer so you know where the decimal point will go.

Example: 37.85 × 9.3

Estimate by rounding to 38 × 9 = 342
This indicates that the answer will be about 342.

Remove the decimal point and calculate:
3785 × 93 = 352 005

Put the decimal point in the answer: 352.005

So, 37.85 × 9.3 = 352.005

Remember

When solving division problems that have remainders, the remainder can be written as a whole number or as a fraction of the group.

Example: $5473 \div 8 = 684$ r 1 or $684\frac{1}{8}$.

At the end of each Strand, there is an Assessment section that covers all the examples from that Strand. If you have difficulty with any test questions, go back to the relevant page in the Strand and look at the worked examples before doing the test again.

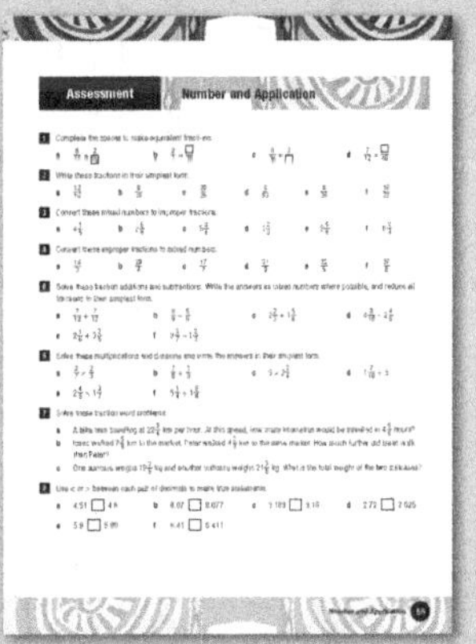

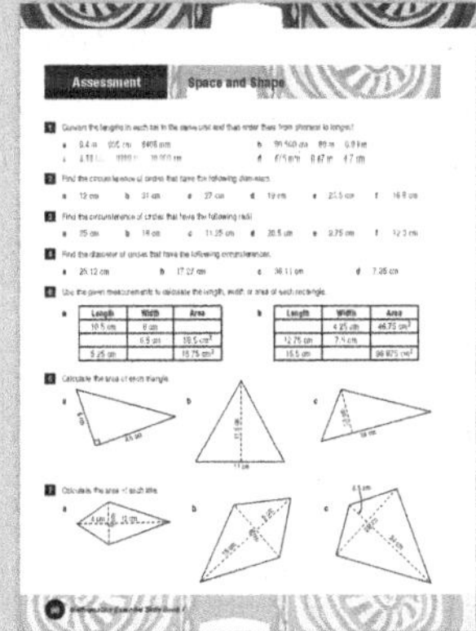

There is an Answers section and a section for Important Facts at the back of the book. The Answers section contains answers to all practice examples and test questions while the Important Facts section contains some of the facts that you will need to refer to throughout the year.

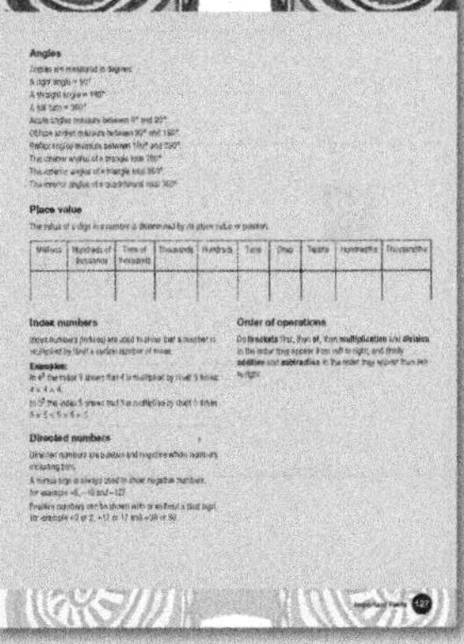

Strand Number and Application

Revise addition with whole numbers

Copy and solve each addition, and then find the total of the answers in each row and column. Add these totals together to find the SUPER TOTAL.

1 36 157 + 26 578	**2** 3 806 5 478 + 4 557	**3** 5 083 + 13 957	
4 46 693 17 808 + 6 523	**5** 516 739 + 283 658	**6** 265 916 + 43 185	
7 72 766 + 408 195	**8** 61 894 276 155 + 53 872	**9** 86 557 4 385 + 33 216	
10 351 544 78 297 + 21 588	**11** 412 607 + 59 018	**12** 75 768 89 345 + 4 263	
13 96 175 28 648 + 7 896	**14** 47 083 + 266 736	**15** 78 299 106 525 + 32 380	
			SUPER TOTAL

Revise subtraction with whole numbers

Copy and subtract at each stage of the subtraction columns until you reach the number at the bottom. Did you get the answer shown for each column?

1	2	3	4
675 148	406 752	916 162	833 057
− 36 759	− 59 023	− 83 655	− 4 885
− 18 596	− 21 518	− 12 033	− 56 172
− 5 077	− 49 996	− 36 558	− 164 983
− 40 819	− 6 583	− 109 439	− 39 506
− 158 725	− 87 525	− 5 167	− 1 852
− 86 487	− 19 569	− 79 826	− 96 464
− 9 776	− 6 709	− 46 225	− 48 591
− 74 512	− 13 086	− 5 778	− 233 115
− 6 329	− 28 485	− 213 166	− 27 682
− 58 437	− 33 127	− 54 921	− 8 476
− 27 085	− 7 993	− 34 557	− 12 038
− 8 544	− 16 259	− 5 648	− 74 593

56879

229189

64700

Revise multiplication with whole numbers

1 Solve these multiplications.

a 2613 × 4
b 1675 × 7
c 4862 × 5
d 3597 × 8
e 6448 × 6

f 16 738 × 3
g 54 066 × 9
h 27 514 × 6
i 72 825 × 4
j 39 483 × 7

k 48 772 × 8
l 65 719 × 3
m 34 207 × 5
n 81 668 × 6
o 59 375 × 9

2 Solve these multiplications with two-digit multipliers.

a 7682 × 22
b 5196 × 17
c 9445 × 31
d 3258 × 25
e 6794 × 36

f 48 253 × 16
g 21 098 × 27
h 52 756 × 19
i 17 607 × 33
j 35 549 × 24

k 15 884 × 32
l 31 875 × 45
m 63 267 × 23
n 26 656 × 38
o 43 309 × 42

3 Set out and solve these multiplications.

a 8794 × 9
b 2377 × 6
c 5916 × 8
d 6049 × 5
e 24 870 × 7
f 52 336 × 4
g 38 983 × 6
h 19 748 × 9
i 4859 × 15
j 7662 × 25
k 5408 × 18
l 9263 × 34
m 17 665 × 23
n 21 097 × 32
o 44 558 × 28
p 36 727 × 37

4 Which of these multiplications have products larger than one million?

a 26 868 × 45
b 41 905 × 21
c 68 714 × 13
d 35 133 × 38

Revise multiplication with whole numbers

Play this game with a friend. Each player needs a calculator and a set of counters (each set a different colour).

Take turns to choose two of the numbers in circles and estimate their product. Use a calculator to multiply the two numbers. If the product is on the grid, cover it with a counter. The winner is the first player to cover five numbers in a line in any direction.

49 88 51 24 32 99 75 21

7425	4851	2376	2112	6600
1575	4488	768	2400	5049
8712	1071	2499	4312	1176
2079	1029	2816	1800	3825
3168	1848	1224	672	1568

Revise division with whole numbers

Remember

When solving division problems that have remainders, the remainder can be written as a whole number or as a fraction of the group.

Example: $5473 \div 8 = 684$ r 1 or $684\frac{1}{8}$.

1 Solve these divisions and write any remainders as fractions reduced to their lowest terms.

a	$5\overline{)2783}$	b	$4\overline{)7158}$	c	$8\overline{)3372}$	d	$6\overline{)4915}$	e	$7\overline{)5803}$
f	$9\overline{)47\,232}$	g	$3\overline{)71\,908}$	h	$5\overline{)29\,344}$	i	$7\overline{)30\,286}$	j	$8\overline{)52\,575}$

2 The answers to these divisions are written below the grid. Solve each one and write the letter that matches each answer to discover the name of a place in Papua New Guinea.

E $4\overline{)9274}$	O $7\overline{)6512}$	T $5\overline{)12\,956}$
I $3\overline{)52\,470}$	F $8\overline{)35\,792}$	N $6\overline{)25\,497}$
L $9\overline{)41\,588}$	M $5\overline{)64\,371}$	E $7\overline{)32\,069}$

___	___	___	___	___	___	___	___	___
$2591\frac{1}{5}$	$4581\frac{2}{7}$	$4620\frac{8}{9}$	$2318\frac{1}{2}$	4474	$930\frac{2}{7}$	$12\,874\frac{1}{5}$	17 490	$4249\frac{1}{2}$

3 The answers to these divisions are written below the grid. Solve each one and write the letter that matches each answer to discover the name of a country near Papua New Guinea.

A $7\overline{)6347}$	I $4\overline{)11\,952}$	R $8\overline{)6483}$
L $5\overline{)73\,985}$	A $3\overline{)56\,407}$	S $9\overline{)31\,818}$
U $6\overline{)22\,384}$	T $7\overline{)47\,068}$	A $4\overline{)67\,252}$

___	___	___	___	___	___	___	___	___
$18\,802\frac{1}{3}$	$3730\frac{2}{3}$	$3535\frac{1}{3}$	6724	$810\frac{3}{8}$	16 813	14 797	2988	$906\frac{5}{7}$

Remember

When solving division problems, any remainders can be written as a whole number or as a fraction of the group.

4 Match each division with its answer.

a $19\overline{)4712}$ b $25\overline{)6817}$ c $16\overline{)5073}$ d $32\overline{)5586}$

e $272\frac{17}{25}$ f $174\frac{9}{16}$ g 248 h $317\frac{1}{16}$

5 Solve these divisions.

a $15\overline{)3954}$ b $23\overline{)4827}$ c $35\overline{)7288}$ d $24\overline{)5174}$

e $21\overline{)3522}$ f $18\overline{)4065}$ g $27\overline{)6821}$ h $32\overline{)7156}$

i $16\overline{)25\,944}$ j $25\overline{)37\,068}$ k $22\overline{)45\,362}$ l $19\overline{)76\,258}$

m $31\overline{)47\,141}$ n $42\overline{)92\,647}$ o $36\overline{)81\,558}$ p $17\overline{)54\,239}$

6 Which of these divisions have quotients smaller than 500?

a $14\overline{)3128}$ b $21\overline{)18\,098}$ c $26\overline{)45\,665}$ d $18\overline{)7443}$

e $38\overline{)15\,890}$ f $29\overline{)34\,677}$ g $34\overline{)41\,363}$ h $32\overline{)15\,938}$

7 Which of these divisions have quotients larger than 1500?

a $17\overline{)23\,912}$ b $24\overline{)49\,891}$ c $21\overline{)29\,366}$ d $13\overline{)34\,329}$

e $33\overline{)47\,916}$ f $27\overline{)62\,308}$ g $36\overline{)51\,717}$ h $19\overline{)42\,095}$

8 Solve each division and then find the total of their quotients. Is it more or less than 10 000?

a $25\overline{)10\,375}$ b $28\overline{)7952}$ c $18\overline{)12\,474}$ d $33\overline{)10\,428}$

e $16\overline{)9248}$ f $37\overline{)17\,279}$ g $23\overline{)22\,172}$ h $41\overline{)22\,509}$

i $26\overline{)9204}$ j $44\overline{)10\,692}$ k $29\overline{)14\,152}$ l $35\overline{)25\,375}$

m $17\overline{)14\,518}$ n $27\overline{)14\,472}$ o $48\overline{)31\,824}$ p $21\overline{)16\,464}$

Use the four processes to revise word problems

1 A total of 13 754 people passed through Jacksons International airport during an eight-day period. What was the average number of people per day who passed through the airport?

2 Six cars, each weighing 3178 kilograms, were loaded on to a transporter. What was the total weight of the six cars?

3 A land-owner has three water tanks with the following capacities: 247 874 L, 27 542 L and 172 134 L. What is the total capacity of his three tanks?

4 A water tank contained 195 851 L. If 58 674 L was used to water crops, how much was left in the tank?

5 The owner of a hotel takes K17 856 each day that her hotel is fully booked. How much money will she take if the hotel is fully booked for:

a 7 days? **b** 12 days? **c** 24 days? **d** 30 days?

6 One person owns two hotels that are regularly booked out by tourists. For the month of July, the small hotel took K675 483 while the large hotel took K916 725.

- **a** How much money in total was taken for the month of July?
- **b** What was the difference in the money taken between the large and small hotels?
- **c** If the owner paid K957 815 for costs associated with running the hotels, how much profit did she make for the month of July?

7 A small company bought cars for its staff to drive. It bought three cars at K27 995 each and four cars at K33 750 each. How much did the company spend in total?

8 The land area in square kilometres of six countries is shown in this chart.

Papua New Guinea	462 840	New Zealand	270 467
Australia	7 692 024	Fiji	18 272
Vanuatu	12 189	Japan	377 930

- **a** How much larger is the land area of Australia than Papua New Guinea?
- **b** How much smaller is the land area of Fiji than New Zealand?
- **c** What is the combined land area of Japan and Vanuatu?
- **d** What is the combined land area of Australia and New Zealand?
- **e** How much more land area does Papua New Guinea have than Japan?

9 A wealthy man died and left K874 656 to be shared equally among his 15 relatives. How much money did each relative receive?

7.1.1 Solve problems requiring any of the four operations including mixed numbers

Make equivalent fractions

Help Box

We can create equivalent fractions by dividing the numerator and the denominator by the same number.

$$\frac{4 \div 4}{8 \div 4} = \frac{1}{2}$$

So $\frac{4}{8}$ is the same as $\frac{1}{2}$.

We can also create equivalent fractions by multiplying the numerator and the denominator by the same number.

$$\frac{2 \times 3}{5 \times 3} = \frac{6}{15}$$

So $\frac{2}{5}$ is the same as $\frac{6}{15}$.

1 Which fraction in each row is not equivalent to the other fractions?

a $\frac{1}{2}$ $\frac{4}{8}$ $\frac{11}{20}$ $\frac{5}{10}$
b $\frac{2}{8}$ $\frac{1}{4}$ $\frac{5}{20}$ $\frac{1}{3}$
c $\frac{2}{6}$ $\frac{1}{2}$ $\frac{1}{3}$ $\frac{3}{9}$
d $\frac{75}{100}$ $\frac{3}{4}$ $\frac{7}{12}$ $\frac{9}{12}$
e $\frac{2}{4}$ $\frac{3}{5}$ $\frac{3}{6}$ $\frac{6}{12}$
f $\frac{1}{8}$ $\frac{2}{16}$ $\frac{3}{24}$ $\frac{4}{30}$
g $\frac{3}{10}$ $\frac{1}{5}$ $\frac{2}{10}$ $\frac{5}{25}$
h $\frac{2}{3}$ $\frac{3}{5}$ $\frac{4}{6}$ $\frac{6}{9}$
i $\frac{1}{10}$ $\frac{10}{100}$ $\frac{3}{30}$ $\frac{5}{40}$
j $\frac{2}{4}$ $\frac{9}{16}$ $\frac{3}{6}$ $\frac{7}{14}$
k $\frac{5}{6}$ $\frac{5}{8}$ $\frac{25}{30}$ $\frac{10}{12}$
l $\frac{9}{21}$ $\frac{3}{7}$ $\frac{7}{8}$ $\frac{6}{14}$
m $\frac{7}{10}$ $\frac{3}{5}$ $\frac{15}{25}$ $\frac{60}{100}$
n $\frac{4}{9}$ $\frac{20}{45}$ $\frac{12}{27}$ $\frac{8}{16}$

2 Which fraction in each row is equivalent to the fraction on the left?

a $\frac{3}{4}$ — $\frac{2}{3}$ $\frac{9}{12}$ $\frac{7}{10}$
b $\frac{2}{3}$ — $\frac{3}{9}$ $\frac{4}{8}$ $\frac{8}{12}$
c $\frac{1}{2}$ — $\frac{7}{8}$ $\frac{6}{12}$ $\frac{3}{10}$
d $\frac{1}{3}$ — $\frac{2}{6}$ $\frac{4}{9}$ $\frac{5}{8}$
e $\frac{1}{4}$ — $\frac{4}{16}$ $\frac{8}{24}$ $\frac{2}{5}$
f $\frac{4}{5}$ — $\frac{6}{8}$ $\frac{8}{10}$ $\frac{2}{3}$
g $\frac{3}{10}$ — $\frac{6}{10}$ $\frac{3}{8}$ $\frac{30}{100}$
h $\frac{5}{6}$ — $\frac{8}{10}$ $\frac{10}{12}$ $\frac{20}{30}$
i $\frac{3}{8}$ — $\frac{9}{24}$ $\frac{1}{3}$ $\frac{6}{12}$
j $\frac{2}{5}$ — $\frac{4}{6}$ $\frac{8}{18}$ $\frac{4}{10}$
k $\frac{1}{8}$ — $\frac{2}{16}$ $\frac{3}{20}$ $\frac{1}{16}$
l $\frac{7}{10}$ — $\frac{15}{20}$ $\frac{70}{100}$ $\frac{7}{14}$
m $\frac{4}{9}$ — $\frac{12}{25}$ $\frac{2}{5}$ $\frac{12}{27}$
n $\frac{3}{7}$ — $\frac{15}{35}$ $\frac{12}{25}$ $\frac{6}{15}$
o $\frac{5}{8}$ — $\frac{5}{16}$ $\frac{1}{2}$ $\frac{15}{24}$

3 Complete the spaces to make an equivalent fraction for each of these fractions.

a $\frac{1}{4} = \frac{\square}{12}$
b $\frac{1}{2} = \frac{\square}{10}$
c $\frac{2}{3} = \frac{8}{\square}$
d $\frac{3}{5} = \frac{30}{\square}$
e $\frac{7}{8} = \frac{\square}{32}$
f $\frac{3}{4} = \frac{9}{\square}$
g $\frac{1}{6} = \frac{6}{\square}$
h $\frac{7}{10} = \frac{\square}{50}$
i $\frac{1}{8} = \frac{\square}{64}$
j $\frac{1}{5} = \frac{5}{\square}$
k $\frac{3}{4} = \frac{\square}{40}$
l $\frac{3}{10} = \frac{6}{\square}$
m $\frac{4}{7} = \frac{8}{\square}$
n $\frac{5}{6} = \frac{\square}{30}$
o $\frac{4}{9} = \frac{8}{\square}$
p $\frac{3}{8} = \frac{9}{\square}$
q $\frac{2}{3} = \frac{6}{\square}$
r $\frac{1}{2} = \frac{\square}{50}$
s $\frac{2}{5} = \frac{\square}{10}$
t $\frac{5}{9} = \frac{15}{\square}$
u $\frac{3}{7} = \frac{\square}{21}$
v $\frac{7}{8} = \frac{42}{\square}$
w $\frac{5}{7} = \frac{\square}{35}$
x $\frac{11}{12} = \frac{33}{\square}$
y $\frac{4}{5} = \frac{\square}{25}$

Remember

To reduce a fraction to its simplest form, divide to find its lowest equivalent fraction.

Two equivalent fractions for $\frac{12}{16}$ are $\frac{6}{8}$ and $\frac{3}{4}$ but $\frac{3}{4}$ is its lowest equivalent fraction because it cannot be divided any further.

4 Write the following fractions in their simplest form.

a	$\frac{6}{9}$	b	$\frac{12}{18}$	c	$\frac{6}{24}$	d	$\frac{25}{30}$	e	$\frac{10}{16}$	f	$\frac{6}{15}$
g	$\frac{5}{25}$	h	$\frac{8}{14}$	i	$\frac{6}{10}$	j	$\frac{20}{35}$	k	$\frac{14}{21}$	l	$\frac{15}{27}$
m	$\frac{10}{45}$	n	$\frac{24}{40}$	o	$\frac{15}{18}$	p	$\frac{9}{36}$	q	$\frac{7}{49}$	r	$\frac{16}{64}$
s	$\frac{36}{48}$	t	$\frac{27}{45}$	u	$\frac{12}{32}$	v	$\frac{18}{30}$	w	$\frac{9}{15}$	x	$\frac{25}{75}$

Convert between mixed numbers and improper fractions

Help Box

To convert an improper fraction to a mixed number, divide the numerator by the denominator.

Example: $\frac{11}{3} = 11 \div 3 = 3\frac{2}{3}$

To convert a mixed number to an improper fraction, multiply the denominator by the whole number. Add the answer to the numerator of the fraction and place the total over the denominator.

Example: for $3\frac{5}{7}$, multiply 7 by 3 (21) and add 21 to 5 (26).

Place 26 over 7: $\frac{26}{7}$

1 Convert these mixed numbers to improper fractions.

a	$3\frac{1}{3}$	b	$2\frac{2}{5}$	c	$1\frac{3}{4}$	d	$4\frac{1}{2}$	e	$2\frac{7}{8}$	f	$1\frac{5}{6}$
g	$2\frac{3}{8}$	h	$3\frac{4}{5}$	i	$1\frac{2}{7}$	j	$5\frac{2}{3}$	k	$1\frac{7}{10}$	l	$4\frac{1}{4}$
m	$5\frac{4}{7}$	n	$1\frac{7}{8}$	o	$2\frac{2}{9}$	p	$6\frac{1}{3}$	q	$3\frac{5}{11}$	r	$2\frac{3}{10}$
s	$7\frac{1}{2}$	t	$4\frac{1}{6}$	u	$3\frac{5}{8}$	v	$8\frac{3}{4}$	w	$1\frac{3}{5}$	x	$10\frac{1}{3}$
y	$3\frac{7}{9}$	z	$5\frac{7}{8}$								

2 How many:

a quarters in 7?
b sixths in $3\frac{5}{6}$?
c sixths in $2\frac{2}{3}$?
d fifths in $1\frac{6}{10}$?
e eighths in 4?
f eighths in $5\frac{1}{4}$?
g thirds in $2\frac{3}{9}$?
h tenths in $1\frac{5}{10}$?
i tenths in $4\frac{1}{5}$?
j fifths in $3\frac{4}{10}$?
k quarters in $5\frac{1}{2}$?
l twelfths in $1\frac{5}{6}$?
m tenths in $2\frac{3}{5}$?
n eighths in $4\frac{7}{8}$?
o eighths in $2\frac{3}{4}$?
p sixths in $5\frac{1}{3}$?
q fifths in $2\frac{8}{10}$?
r thirds in $4\frac{2}{6}$?
s quarters in $3\frac{1}{2}$?
t twelfths in $3\frac{1}{6}$?
u tenths in $5\frac{4}{5}$?
v sixths in $4\frac{2}{3}$?
w quarters in $6\frac{2}{8}$?
x fifths in $1\frac{2}{10}$?

3 Convert these improper fractions to mixed numbers.

a $\frac{14}{3}$
b $\frac{10}{3}$
c $\frac{25}{6}$
d $\frac{15}{7}$
e $\frac{19}{8}$
f $\frac{47}{6}$
g $\frac{29}{7}$
h $\frac{53}{12}$
i $\frac{31}{5}$
j $\frac{40}{9}$
k $\frac{21}{4}$
l $\frac{33}{8}$
m $\frac{17}{6}$
n $\frac{36}{5}$
o $\frac{27}{4}$
p $\frac{18}{7}$
q $\frac{12}{5}$
r $\frac{41}{6}$
s $\frac{29}{5}$
t $\frac{23}{7}$
u $\frac{20}{9}$
v $\frac{32}{3}$
w $\frac{25}{4}$
x $\frac{34}{9}$
y $\frac{22}{5}$
z $\frac{38}{7}$

4 Convert one fraction in each pair so that you can compare them to find the smaller one.

a $\frac{15}{8}$ or $1\frac{1}{2}$
b $2\frac{3}{4}$ or $\frac{13}{4}$
c $\frac{23}{5}$ or $4\frac{1}{10}$
d $1\frac{4}{7}$ or $\frac{10}{7}$
e $\frac{27}{6}$ or $4\frac{3}{4}$
f $3\frac{2}{3}$ or $\frac{14}{3}$
g $\frac{19}{9}$ or $2\frac{1}{3}$
h $3\frac{1}{4}$ or $\frac{23}{8}$
i $\frac{16}{3}$ or $5\frac{1}{6}$
j $2\frac{7}{8}$ or $\frac{21}{8}$
k $\frac{17}{6}$ or $2\frac{1}{2}$
l $1\frac{9}{10}$ or $\frac{7}{5}$
m $\frac{33}{7}$ or $4\frac{6}{7}$
n $5\frac{2}{5}$ or $\frac{11}{2}$
o $\frac{25}{4}$ or $6\frac{3}{4}$
p $1\frac{3}{8}$ or $\frac{3}{2}$
q $\frac{19}{5}$ or $3\frac{1}{10}$
r $2\frac{1}{3}$ or $\frac{13}{6}$
s $\frac{31}{8}$ or $3\frac{3}{4}$
t $6\frac{1}{2}$ or $\frac{25}{4}$
u $\frac{40}{9}$ or $4\frac{2}{9}$
v $3\frac{5}{11}$ or $\frac{40}{11}$
w $\frac{16}{5}$ or $3\frac{3}{10}$
x $7\frac{5}{6}$ or $\frac{23}{3}$

Add fractions including mixed numbers

Help Box

To add mixed numbers, first add the whole numbers.

Example: for $3\frac{2}{5} + 1\frac{2}{3}$, $3 + 1 = 4$

The addition now becomes $4 + \frac{2}{5} + \frac{2}{3}$.

Convert $\frac{2}{5}$ and $\frac{2}{3}$ to equivalent fractions with a common denominator of 15 ($\frac{6}{15} + \frac{10}{15}$).

The addition now becomes $4 + \frac{6}{15} + \frac{10}{15} = 4 + \frac{16}{15}$, which can be simplified to $5\frac{1}{15}$.

1 Calculate answers to these fraction additions. Write your answers as mixed numbers where possible and reduce all fractions to their simplest form.

a $\frac{5}{6} + \frac{5}{6}$ **b** $\frac{7}{10} + \frac{2}{5}$ **c** $1\frac{4}{5} + \frac{4}{5}$ **d** $\frac{5}{7} + 2\frac{6}{7}$

e $3\frac{1}{3} + \frac{5}{9}$ **f** $\frac{3}{10} + \frac{11}{100}$ **g** $2\frac{4}{5} + 1\frac{6}{10}$ **h** $1\frac{5}{6} + \frac{3}{12}$

i $2\frac{3}{4} + \frac{7}{12}$ **j** $3\frac{2}{5} + \frac{9}{10}$ **k** $1\frac{3}{4} + \frac{5}{8}$ **l** $3\frac{1}{2} + 1\frac{1}{8}$

m $1\frac{9}{10} + \frac{4}{5}$ **n** $2\frac{5}{6} + \frac{2}{3}$ **o** $\frac{5}{12} + 1\frac{5}{6}$ **p** $3\frac{7}{10} + 2\frac{1}{5}$

q $4\frac{1}{4} + 2\frac{7}{8}$ **r** $\frac{4}{9} + 2\frac{2}{3}$ **s** $\frac{3}{8} + 5\frac{1}{2}$ **t** $2\frac{2}{3} + 4\frac{1}{12}$

u $1\frac{3}{4} + 4\frac{5}{12}$ **v** $6\frac{9}{10} + 4\frac{4}{5}$ **w** $3\frac{11}{15} + 4\frac{1}{3}$ **x** $2\frac{17}{24} + \frac{7}{8}$

2 Add each group of fractions or mixed numbers and write the answers in their simplest form.

a $2\frac{4}{5} + \frac{1}{5} + 1\frac{7}{10}$ **b** $3\frac{5}{6} + 2\frac{2}{3} + 1\frac{5}{6}$ **c** $1\frac{3}{4} + 1\frac{7}{8} + 3\frac{1}{8}$

d $\frac{1}{10} + 3\frac{1}{2} + 1\frac{3}{5}$ **e** $\frac{7}{12} + 4\frac{1}{3} + 2\frac{5}{12}$ **f** $2\frac{5}{9} + \frac{1}{3} + 1\frac{7}{9}$

g $5\frac{1}{2} + 3\frac{5}{8} + 1\frac{3}{4}$ **h** $1\frac{2}{5} + \frac{3}{10} + 4\frac{9}{10}$ **i** $2\frac{7}{10} + 2\frac{2}{5} + 1\frac{3}{5}$

j $3\frac{1}{6} + \frac{11}{12} + 2\frac{2}{3}$ **k** $2\frac{7}{8} + 1\frac{1}{4} + 7\frac{1}{2}$ **l** $1\frac{4}{9} + 4\frac{1}{2} + 2\frac{1}{3}$

m $\frac{7}{12} + 4\frac{2}{3} + 2\frac{1}{6}$ **n** $3\frac{5}{8} + 6\frac{3}{4} + \frac{11}{16}$ **o** $\frac{8}{15} + 2\frac{2}{3} + 3\frac{3}{5}$

p $5\frac{1}{8} + 2\frac{1}{4} + 3\frac{3}{8}$ **q** $3\frac{1}{2} + 1\frac{11}{12} + 2\frac{5}{6}$ **r** $\frac{9}{10} + 2\frac{3}{10} + 1\frac{4}{5}$

s $6\frac{2}{5} + \frac{13}{20} + 1\frac{7}{10}$ **t** $7\frac{3}{4} + 5\frac{5}{8} + \frac{9}{16}$ **u** $3\frac{4}{7} + 2\frac{1}{3} + \frac{5}{7}$

v $\frac{7}{9} + 5\frac{2}{3} + 3\frac{5}{9}$ **w** $\frac{13}{15} + 1\frac{4}{5} + \frac{1}{3}$ **x** $8\frac{1}{4} + 5\frac{5}{12} + \frac{7}{24}$

3 Add these fractions and mixed numbers to see how many rocks you can step on. Write all answers in their simplest form.

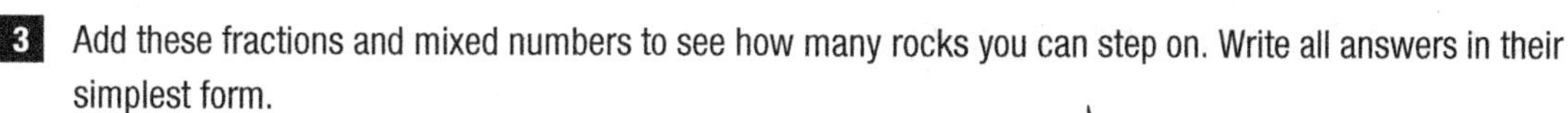

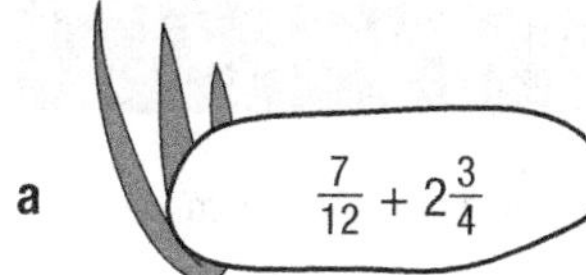

a $\frac{7}{12} + 2\frac{3}{4}$

b $1\frac{2}{3} + 2\frac{5}{6}$

c $4\frac{1}{3} + 1\frac{7}{9}$

d $3\frac{5}{9} + 4\frac{2}{3}$

e $2\frac{1}{4} + \frac{11}{12}$

f $\frac{5}{7} + 1\frac{2}{3}$

g $6\frac{2}{3} + 4\frac{7}{12}$

h $1\frac{4}{7} + 4\frac{1}{2}$

i $2\frac{5}{11} + \frac{2}{3}$

j $1\frac{7}{10} + 3\frac{2}{5}$

k $4\frac{1}{6} + 1\frac{5}{12}$

l $7\frac{1}{3} + 2\frac{1}{12}$

m $4\frac{7}{8} + 7\frac{1}{6}$

n $2\frac{6}{7} + 1\frac{3}{5}$

o $\frac{8}{9} + 3\frac{1}{5}$

p $7\frac{3}{4} + 3\frac{4}{7}$

q $1\frac{6}{11} + 2\frac{1}{4}$

r $\frac{5}{8} + 3\frac{1}{3}$

How many rocks did you step on?

Subtract fractions including mixed numbers

Help Box

To subtract mixed numbers, first convert both numbers to improper fractions.

Example: for $5\frac{1}{5} - 2\frac{3}{4}$, the subtraction now becomes $\frac{26}{5} - \frac{11}{4}$.

Convert $\frac{26}{5}$ and $\frac{11}{4}$ to equivalent fractions with a common denominator of 20.

The subtraction now becomes $\frac{104}{20} - \frac{55}{20} = \frac{49}{20}$, which can be simplified to $2\frac{9}{20}$.

1 Calculate answers to these fraction subtractions. Write your answers as mixed numbers where possible and reduce all fractions to their simplest form.

a	$\frac{7}{8} - \frac{3}{8}$	**b**	$\frac{11}{12} - \frac{6}{12}$	**c**	$1\frac{2}{11} - \frac{4}{11}$	**d**	$\frac{17}{20} - \frac{3}{10}$
e	$2\frac{8}{15} - \frac{4}{5}$	**f**	$1\frac{9}{10} - \frac{2}{5}$	**g**	$2\frac{5}{8} - \frac{1}{4}$	**h**	$4\frac{1}{6} - \frac{2}{3}$
i	$3\frac{4}{15} - 1\frac{3}{5}$	**j**	$1\frac{31}{100} - \frac{7}{10}$	**k**	$5\frac{1}{6} - 2\frac{1}{3}$	**l**	$4\frac{1}{3} - 2\frac{5}{6}$
m	$3\frac{7}{10} - \frac{4}{5}$	**n**	$4\frac{1}{12} - 1\frac{1}{6}$	**o**	$4\frac{1}{8} - 1\frac{3}{4}$	**p**	$2\frac{4}{9} - 1\frac{2}{3}$
q	$7\frac{1}{2} - 4\frac{7}{10}$	**r**	$5\frac{3}{4} - 3\frac{11}{12}$	**s**	$4\frac{1}{5} - 1\frac{1}{3}$	**t**	$3\frac{4}{7} - 1\frac{1}{2}$
u	$8\frac{2}{7} - 5\frac{2}{3}$	**v**	$5\frac{1}{2} - 2\frac{4}{5}$	**w**	$4\frac{3}{10} - 1\frac{2}{5}$	**x**	$6\frac{2}{7} - 3\frac{3}{5}$

2 The answers to these subtractions are written below the grid. Work out each one and write the letter that matches each answer to discover the name of a country.

N $2\frac{3}{8} - 1\frac{2}{3}$	**E** $5\frac{1}{5} - 2\frac{7}{10}$	**A** $1\frac{5}{12} - \frac{5}{6}$
I $4\frac{1}{4} - 1\frac{3}{5}$	**S** $3\frac{2}{7} - 1\frac{3}{4}$	**N** $2\frac{4}{9} - 1\frac{1}{2}$
O $3\frac{1}{10} - \frac{4}{7}$	**D** $6\frac{2}{3} - 4\frac{5}{8}$	**I** $4\frac{3}{4} - 1\frac{1}{3}$

___	___	___	___	___	___	___	___	___
$3\frac{5}{12}$	$\frac{17}{18}$	$2\frac{1}{24}$	$2\frac{37}{70}$	$\frac{17}{24}$	$2\frac{1}{2}$	$1\frac{15}{28}$	$2\frac{13}{20}$	$\frac{7}{12}$

3 Subtract these fractions and mixed numbers to see how many ducks you can catch. Write all answers in their simplest form.

a  $3\frac{5}{12} - 2\frac{3}{4}$

b  $5\frac{3}{7} - 2\frac{1}{3}$

c  $2\frac{1}{10} - \frac{5}{8}$

d  $4\frac{1}{9} - 2\frac{5}{6}$

e  $2\frac{6}{7} - \frac{1}{2}$

f  $5\frac{2}{3} - 2\frac{11}{12}$

g  $6\frac{1}{2} - 3\frac{2}{3}$

h $4\frac{1}{4} - 1\frac{7}{12}$

i $3\frac{2}{5} - 1\frac{9}{10}$

j  $2\frac{3}{8} - \frac{3}{10}$

k  $3\frac{1}{6} - 1\frac{1}{8}$

l  $4\frac{1}{3} - 2\frac{5}{12}$

m  $5\frac{4}{5} - 3\frac{1}{4}$

n  $2\frac{8}{9} - 1\frac{1}{2}$

o  $7\frac{1}{5} - 2\frac{4}{7}$

p  $5\frac{1}{4} - 2\frac{7}{12}$

q  $3\frac{2}{11} - 1\frac{3}{5}$

r  $4\frac{2}{7} - 2\frac{1}{3}$

How many ducks did you catch?

Multiply fractions including mixed numbers

Help Box

To multiply fractions, first multiply both the numerators and the denominators.

Example: for $\frac{3}{8} \times \frac{4}{5}$, multiply (3 × 4) and (8 × 5).

The multiplication now becomes $\frac{12}{40}$, which can be simplified to $\frac{3}{10}$.

To multiply mixed numbers, first change the mixed numbers to improper fractions.

Example: for $1\frac{2}{3} \times 2\frac{3}{4}$, change to $\frac{5}{3} \times \frac{11}{4}$.

The multiplication now becomes $\frac{55}{12}$, which can be simplified to $4\frac{7}{12}$.

1 Calculate each multiplication and match it with the correct answer.

a $\frac{1}{7} \times \frac{2}{9}$ b $\frac{3}{5} \times \frac{4}{11}$ c $\frac{2}{3} \times \frac{7}{10}$ d $\frac{5}{8} \times \frac{5}{6}$ e $\frac{1}{3} \times \frac{3}{7}$

f $\frac{3}{4} \times \frac{4}{5}$ g $\frac{7}{8} \times \frac{2}{3}$ h $\frac{5}{9} \times \frac{2}{7}$ i $\frac{3}{10} \times \frac{1}{2}$ j $\frac{2}{3} \times \frac{2}{3}$

Answers: $\frac{1}{7}$ $\frac{10}{63}$ $\frac{12}{55}$ $\frac{3}{5}$ $\frac{3}{20}$ $\frac{2}{63}$ $\frac{4}{9}$ $\frac{25}{48}$ $\frac{7}{12}$ $\frac{7}{15}$

2 Calculate and write the answer in its simplest form.

a $1\frac{3}{4} \times 2\frac{1}{2}$ b $2\frac{3}{5} \times 1\frac{3}{8}$ c $3\frac{1}{4} \times 1\frac{2}{3}$ d $2\frac{7}{8} \times 2\frac{1}{5}$

e $2\frac{1}{3} \times 1\frac{5}{7}$ f $1\frac{9}{10} \times 3\frac{1}{2}$ g $2\frac{4}{5} \times 4\frac{2}{3}$ h $3\frac{2}{7} \times 1\frac{8}{9}$

i $4\frac{1}{4} \times 2\frac{1}{6}$ j $1\frac{8}{11} \times 1\frac{2}{5}$ k $4\frac{4}{7} \times 2\frac{1}{3}$ l $2\frac{1}{7} \times 2\frac{5}{8}$

Remember

1. When multiplying a fraction or mixed number by a whole number, write the whole number as a fraction over 1.
   Example: $5 \times 2\frac{3}{4} = \frac{5}{1} \times \frac{11}{4}$
2. Calculations involving the word 'of', such as $\frac{2}{3}$ of 18 and $\frac{5}{8}$ of $2\frac{3}{4}$, can be solved by multiplication.
   Examples: $\frac{2}{3}$ of $18 = \frac{2}{3} \times \frac{18}{1} = \frac{36}{3} = 12$; $\frac{5}{8}$ of $2\frac{3}{4} = \frac{5}{8} \times \frac{11}{4} = \frac{55}{32} = 1\frac{23}{32}$

3 Calculate and write the answer in its simplest form.

a $8 \times 1\frac{2}{3}$ b $3\frac{4}{5} \times 6$ c $\frac{7}{8} \times 1\frac{2}{5}$ d $2\frac{7}{8} \times \frac{1}{2}$ e $\frac{6}{7} \times 9$

f $4\frac{1}{3} \times \frac{5}{7}$ g $\frac{9}{10} \times 1\frac{3}{4}$ h $\frac{4}{9} \times 7$ i $3\frac{1}{6} \times \frac{1}{4}$ j $5 \times 4\frac{7}{9}$

4 Calculate and write the answer in its simplest form.

a $\frac{3}{4}$ of $3\frac{4}{5}$ b $\frac{1}{3}$ of $2\frac{7}{8}$ c $\frac{5}{6}$ of 24 d $\frac{4}{7}$ of $2\frac{2}{3}$ e $\frac{8}{9}$ of 72

f $\frac{3}{5}$ of 25 g $\frac{2}{3}$ of $1\frac{5}{9}$ h $\frac{5}{8}$ of $2\frac{1}{4}$ i $\frac{4}{9}$ of $3\frac{1}{7}$ j $\frac{7}{8}$ of 56

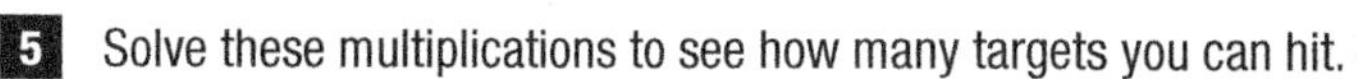

5 Solve these multiplications to see how many targets you can hit.

a 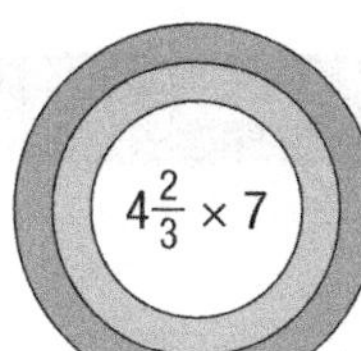 $4\frac{2}{3} \times 7$

b 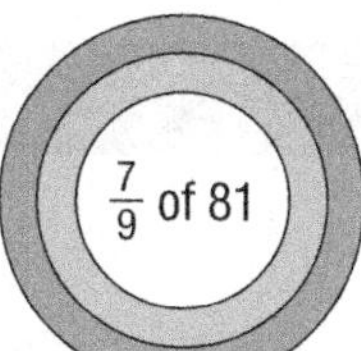 $\frac{7}{9}$ of 81

c 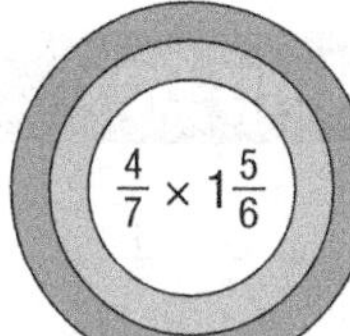 $\frac{4}{7} \times 1\frac{5}{6}$

d 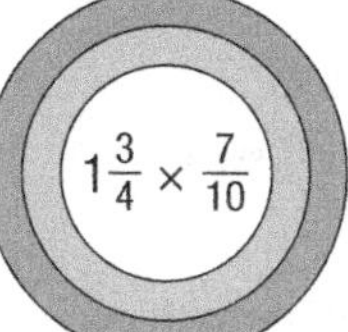 $1\frac{3}{4} \times \frac{7}{10}$

e $2\frac{1}{3} \times 4\frac{2}{7}$

f 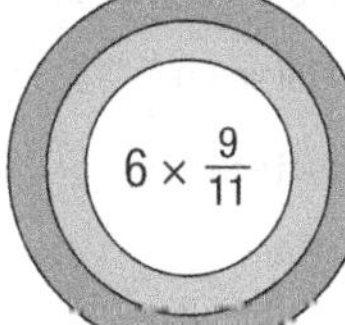 $6 \times \frac{9}{11}$

g $\frac{5}{8}$ of $7\frac{1}{2}$

h $\frac{6}{11} \times \frac{2}{9}$

i $3\frac{4}{7} \times 1\frac{5}{8}$

j $\frac{2}{5} \times 2\frac{7}{8}$

k $\frac{3}{5}$ of $6\frac{3}{4}$

l $3\frac{2}{3} \times 8$

m $\frac{3}{8} \times \frac{5}{12}$

n $1\frac{9}{10} \times 4\frac{1}{4}$

o $3\frac{2}{7} \times 2\frac{1}{5}$

p $\frac{5}{7}$ of 85

q $5 \times 6\frac{2}{9}$

r 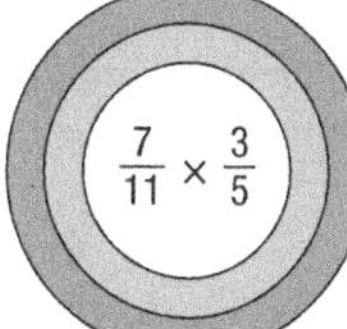 $\frac{7}{11} \times \frac{3}{5}$

s $5\frac{3}{4} \times 2\frac{2}{5}$

t $1\frac{3}{10} \times \frac{7}{12}$

u $\frac{9}{20} \times 8$

How many targets did you hit?

Divide fractions including mixed numbers

Help Box

Example: $\frac{5}{8} \div \frac{3}{5}$

To divide fractions, first invert $\frac{3}{5}$ to become $\frac{5}{3}$ and replace the + symbol with a × symbol.

The division now becomes a multiplication: $\frac{5}{8} \times \frac{5}{3}$.

Calculate the answer $\frac{25}{24}$, which can be simplified to $1\frac{1}{24}$.

Example: $3\frac{1}{3} \div 1\frac{2}{7}$

To divide mixed numbers, first change the mixed numbers to improper fractions.

The division now becomes $\frac{10}{3} \div \frac{9}{7}$.

Invert the divisor and replace the ÷ symbol with a × symbol.

The division now becomes a multiplication: $\frac{10}{3} \times \frac{7}{9}$.

Calculate the answer $\frac{70}{27}$, which can be simplified to $2\frac{16}{27}$.

1 Calculate each division and write the answer in its simplest form.

a	$\frac{3}{4} \div \frac{1}{2}$	**b**	$\frac{4}{5} \div \frac{1}{4}$	**c**	$\frac{5}{8} \div \frac{1}{3}$	**d**	$\frac{5}{9} \div \frac{2}{5}$	**e**	$\frac{2}{3} \div \frac{5}{6}$
f	$\frac{7}{8} \div \frac{3}{5}$	**g**	$\frac{8}{9} \div \frac{2}{7}$	**h**	$\frac{5}{11} \div \frac{1}{8}$	**i**	$\frac{7}{10} \div \frac{3}{4}$	**j**	$\frac{3}{8} \div \frac{1}{5}$
k	$2\frac{1}{2} \div 1\frac{4}{7}$	**l**	$3\frac{1}{8} \div 1\frac{1}{2}$	**m**	$1\frac{5}{6} \div 2\frac{3}{4}$	**n**	$4\frac{3}{10} \div 2\frac{1}{5}$	**o**	$2\frac{7}{8} \div 1\frac{1}{4}$
p	$5\frac{6}{7} \div \frac{5}{8}$	**q**	$1\frac{9}{10} \div \frac{4}{5}$	**r**	$2\frac{2}{5} \div \frac{2}{3}$	**s**	$\frac{2}{9} \div 3\frac{1}{8}$	**t**	$\frac{7}{8} \div 2\frac{1}{2}$

Remember

Where the divisor or dividend is a whole number, write the whole number as a fraction over 1. Example: $3 \div \frac{5}{7} = \frac{3}{1} \div \frac{5}{7}$

2 Calculate and write the answer in its simplest form.

a	$\frac{7}{8} \div 5$	**b**	$8 \div \frac{3}{4}$	**c**	$1\frac{3}{5} \div 4$	**d**	$3 \div 2\frac{1}{4}$	**e**	$4\frac{2}{9} \div 7$
f	$10 \div \frac{4}{5}$	**g**	$1\frac{7}{9} \div 6$	**h**	$2 \div 1\frac{2}{3}$	**i**	$\frac{8}{11} \div 9$	**j**	$5 \div 2\frac{1}{5}$
k	$3\frac{1}{2} \div 8$	**l**	$12 \div 1\frac{5}{6}$	**m**	$2\frac{9}{10} \div 3$	**n**	$7 \div 2\frac{4}{5}$	**o**	$\frac{7}{12} \div 4$

3 Use division to find the missing fraction or mixed number. Write all answers in their simplest form.

a	$2\frac{1}{3} \times \square = \frac{7}{8}$	**b**	$\frac{4}{5} \times \square = 1\frac{3}{4}$	**c**	$1\frac{5}{9} \times \square = 2\frac{3}{5}$	**d**	$\square \times \frac{9}{10} = 3\frac{2}{3}$
e	$\square \times 2\frac{1}{4} = 5\frac{5}{8}$	**f**	$\square \times 3\frac{1}{9} = 1\frac{1}{2}$	**g**	$5\frac{1}{6} \times \square = 2\frac{3}{7}$	**h**	$3\frac{2}{5} \times \square = 7\frac{1}{10}$
i	$\frac{7}{12} \times \square = \frac{2}{9}$	**j**	$\square \times 1\frac{2}{3} = 1\frac{3}{4}$	**k**	$\square \times 4\frac{7}{8} = 6\frac{5}{12}$	**l**	$\square \times \frac{7}{9} = 2\frac{1}{3}$

4 Solve these divisions to see how many touchdowns you can score.

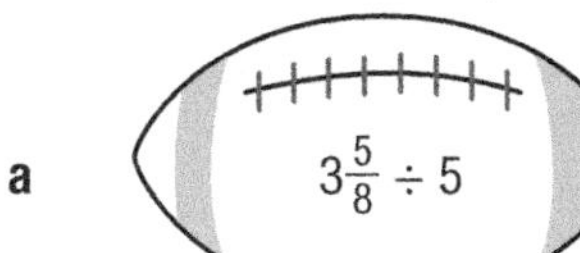

a $3\frac{5}{8} \div 5$

b $2\frac{3}{4} \div \frac{2}{3}$

c $1\frac{3}{7} \div \frac{3}{4}$

d $\frac{8}{11} \div \frac{1}{4}$

e $9 \div 1\frac{4}{5}$

f $3\frac{1}{2} \div 1\frac{7}{10}$

g $\frac{5}{9} \div 2\frac{2}{3}$

h $4\frac{1}{8} \div 2\frac{1}{6}$

i $5\frac{1}{7} \div \frac{4}{9}$

j $7 \div 4\frac{1}{3}$

k $7\frac{2}{5} \div \frac{5}{8}$

l $2\frac{1}{9} \div 4$

m $6\frac{6}{7} \div 4\frac{3}{5}$

n $\frac{10}{11} \div 5\frac{1}{2}$

o $4\frac{5}{7} \div 3\frac{2}{5}$

p $\frac{7}{12} \div 1\frac{3}{10}$

q $1\frac{9}{20} \div 5\frac{1}{5}$

r $12 \div 4\frac{1}{12}$

s $8\frac{3}{4} \div 1\frac{5}{6}$

t $5\frac{11}{15} \div \frac{3}{5}$

u $\frac{9}{16} \div 1\frac{1}{7}$

How many touchdowns did you score?

Solve fraction word problems

Decide what process to use and then solve each word problem.

1 Kali has $3\frac{2}{3}$ L of orange juice. How many $\frac{1}{4}$ L cups can she fill?

2 Emily made seven bottles of cordial. Each bottle contained $\frac{7}{8}$ L. How much cordial did she make in total?

3 Samuel bought $4\frac{1}{3}$ kg of minced meat. He used $2\frac{3}{8}$ kg to make hamburgers. How much minced meat is left?

4 Lydia has $3\frac{1}{3}$ square metres of garden. She plants beans in $\frac{2}{5}$ of her garden. How many square metres is planted with beans?

5 A boat was travelling at $15\frac{3}{4}$ km per hour. How many kilometres are travelled in:

a $\frac{1}{2}$ an hour? **b** $\frac{1}{4}$ of an hour? **c** $5\frac{3}{10}$ hours?

6 Grace bought $7\frac{5}{8}$ metres of fabric. She used $4\frac{3}{5}$ metres to make curtains. How much fabric is left?

7 Sharon uses $3\frac{1}{2}$ cups of vinegar in her salad dressing recipe. How much vinegar would Sharon use to make $\frac{1}{3}$ of her recipe?

8 Marie has $5\frac{3}{8}$ kg of rice. How many bags, each containing $\frac{3}{4}$ kg, can she fill?

9 Yerema travels $8\frac{2}{3}$ kilometres each day. How far does she travel in:

a 7 days? **b** 4 days? **c** 11 days? **d** 8 days?

10 Peter's truck gets him $7\frac{2}{3}$ kilometres per litre of fuel. If Peter's tank is empty and he puts in $5\frac{1}{2}$ litres, how far can he go with the truck?

11 Isaac walked $6\frac{1}{5}$ kilometres on Saturday and $4\frac{5}{8}$ kilometres on Sunday.

a How many kilometres did he walk on the two days?

b How much further did he walk on Saturday than Sunday?

12 A parcel weighs $1\frac{4}{5}$ kg. Another parcel weighs $2\frac{3}{4}$ kg. What is the combined weight of the two parcels?

13 While making bread, Jay used $7\frac{2}{3}$ cups of white flour and $1\frac{5}{8}$ cups of wholemeal flour. How many cups of flour did Jay use altogether?

7.1.2 Use decimals in solving problems set in familiar contexts

Compare and order decimals

1 Use < or > between each pair of decimals to make true statements.

a	3.82 ☐ 3.9	**b**	0.7 ☐ 0.07	**c**	2.615 ☐ 2.62	**d**	5.04 ☐ 5.004
e	4.3 ☐ 4.279	**f**	8.12 ☐ 8.2	**g**	6.84 ☐ 6.834	**h**	7.206 ☐ 7.21
i	0.822 ☐ 0.83	**j**	1.65 ☐ 1.605	**k**	9.447 ☐ 9.45	**l**	3.93 ☐ 3.933
m	2.55 ☐ 2.6	**n**	4.366 ☐ 4.37	**o**	8.3 ☐ 8.299	**p**	6.854 ☐ 6.85
q	1.09 ☐ 1.1	**r**	5.713 ☐ 5.72	**s**	3.5 ☐ 3.06	**t**	4.885 ☐ 4.89
u	7.937 ☐ 7.94	**v**	0.78 ☐ 0.775	**w**	2.048 ☐ 2.05	**x**	9.6 ☐ 9.161

2 Which of the following numbers are more than 8.05?

8.049 8.015 8.005 8.02 8.051 8.09 8.2 8.08 8.055 8.06 8.1 8.045 8.5 8.053 8.15

3 Which of the following numbers are less than 5.7?

5.75 5.65 5.08 5.17 5.99 5.57 5.69 5.67 5.8 5.07 5.008 5.725 5.775 5.1 5.073

4 Which of the following numbers come between 6.08 and 6.3?

6.25 6.31 6.119 6.018 6.206 6.24 6.1 6.099 6.03 6.085 6.46 6.058 6.17 6.211 6.09

5 Which of the following numbers come between 2.97 and 3.16?

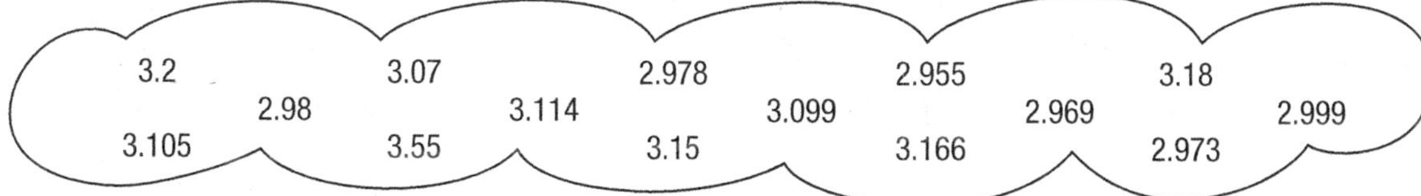

6 Order each set of numbers from smallest to largest.

- **a** 5.67 5.067 5.7 5.66 5.07 5.5 5.06 5.17
- **b** 2.38 2.03 2.88 2.138 2.025 2.83 2.3 2.25
- **c** 9.1 9.115 9.9 9.09 9.189 9.017 9.82 9.875
- **d** 0.45 0.54 0.455 0.5 0.04 0.499 0.55 0.05
- **e** 4.9 4.89 4.99 4.989 4.09 4.8 4.98 4.85
- **f** 7.32 7.5 7.03 7.23 7.032 7.55 7.3 7.035
- **g** 1.067 1.7 1.06 1.67 1.675 1.75 1.68 1.6
- **h** 3.15 3.155 3.1 3.5 3.51 3.015 3.25 3.01
- **i** 6.44 6.4 6.04 6.45 6.045 6.004 6.5 6.05
- **j** 8.2 8.26 8.16 8.19 8.129 8.25 8.026 8.13
- **k** 2.775 2.76 2.77 2.077 2.07 2.6 2.8 2.7
- **l** 0.33 0.4 0.335 0.3 0.035 0.39 0.345 0.34

Round decimals

Help Box

When rounding a decimal number to a specific number of decimal places, look at the value of the next decimal place.

Example 1: Round 6.853 to one decimal place.

Look at the second decimal place (5) and if it is 5 or more, increase the first decimal place (8) by one. So, 6.853 rounded to one decimal place is 6.9

Example 2: Round 0.073482 to three decimal places.

Look at the fourth decimal place (4) and if it is less than 5 leave the third decimal place as it is (3). So, 0.073482 rounded to three decimal places is 0.073

1 Round these numbers to one decimal place (the nearest tenth).

a	3.58	**b**	8.733	**c**	0.2614	**d**	6.425	**e**	10.07	**f**	5.381
g	11.622	**h**	4.4392	**i**	12.87	**j**	2.154	**k**	9.3755	**l**	7.008

2 Round these numbers to two decimal places (the nearest hundredth).

a	7.536	**b**	0.8187	**c**	3.666	**d**	15.323	**e**	11.0644	**f**	2.915
g	8.1538	**h**	5.296	**i**	10.8412	**j**	6.0875	**k**	4.7533	**l**	1.5567

3 Round these numbers to three decimal places (the nearest thousandth).

a	15.7146	**b**	5.8644	**c**	8.0755	**d**	0.4563	**e**	12.2249	**f**	2.3176
g	6.5807	**h**	3.4992	**i**	19.99967	**j**	13.76351	**k**	7.08005	**l**	4.2121

4 Round these numbers to zero decimal places (the nearest whole number).

a	24.537	**b**	16.8186	**c**	9.254	**d**	12.06	**e**	8.7555	**f**	4.393
g	15.954	**h**	27.8	**i**	36.708	**j**	21.115	**k**	43.2714	**l**	19.787

5 Copy and complete the following table.

	Rounded to three decimal places	Rounded to one decimal place	Rounded to nearest whole number	Rounded to two decimal places
a	4.376			
b	9.105			
c	16.837			
d	10.548			
e	7.294			
f	25.616			
g	0.774			
h	5.441			
i	8.053			
j	12.986			
k	30.289			
l	3.008			
m	1.766			
n	6.251			
o	11.642			

Round decimals to estimate answers

1 Use rounding to estimate whole number answers to the following questions. Show the whole numbers that you used to work out your estimates. Do not work out the exact answers.

a	$4.8 + 11.9$	**b**	$12.1 - 5.8$	**c**	6.3×5.2	**d**	$15.2 \div 3.1$
e	$14.7 + 9.2$	**f**	$25.2 - 12.4$	**g**	10.7×8.9	**h**	$23.8 \div 6.2$
i	$25.15 + 16.97$	**j**	$33.88 - 10.96$	**k**	4.13×7.98	**l**	$36.23 \div 4.07$
m	$18.79 + 17.06$	**n**	$41.21 - 15.87$	**o**	11.94×8.16	**p**	$26.99 \div 2.95$
q	$32.77 + 25.17$	**r**	$29.89 - 15.09$	**s**	20.18×4.89	**t**	$40.25 \div 9.86$
u	$26.22 + 7.887$	**v**	$34.013 - 7.987$	**w**	12.931×3.106	**x**	$31.956 \div 8.114$

2 Use rounding to help you choose the correct answer for each of these calculations.

a	29.89 + 13.06 + 20.9 =	63.85	70.6	59.85	69.15
b	15.04 + 25.2 + 17.88 =	50.62	48.12	58.12	65.52
c	85.13 – 30.007 =	60.03	55.123	52.3	59.233
d	149.87 – 49.859 =	100.011	105.11	110.011	90.11
e	25.122 × 5.998 =	120.682	155.982	145.12	150.68
f	9.875 × 5.025 =	55.62	40.02	49.62	45.52
g	60.3 ÷ 2.99 =	30.167	25.7	15.17	20.167
h	17.865 ÷ 5.97 =	3.59	2.99	2.59	3.99
i	225.75 + 135.9 =	350.65	361.65	300.05	371.65
j	67.98 + 30.03 + 29.89 =	127.9	130.9	120.9	124.9
k	216.25 – 169.95 =	40.3	56.3	49.3	46.3
l	115.75 – 24.983 =	85.7	90.767	87.767	94.67
m	14.886 × 5.015 =	71.53	70.653	78.03	74.653
n	33.018 × 2.96 =	97.733	66.733	70.33	103.33
o	36.018 ÷ 6.002 =	5.95	6.555	6.001	6.901
p	87.96 ÷ 10.995 =	8.99	8.0	9.0	7.59
q	54.7 + 32.89 + 43.65 =	140.24	123.4	119.44	131.24
r	376.45 + 199.78 =	570.3	575.23	576.23	572.33
s	511.05 – 209.97 =	311.08	295.88	301.08	290.38
t	82.125 – 53.066 =	31.129	29.059	25.259	35.599
u	45.87 × 4.03 =	184.856	180.856	204.56	164.56
v	98.99 × 9.99 =	980.91	999.99	988.91	1000.91
w	45.12 ÷ 8.95 =	5.041	6.041	5.999	6.999
x	63.85 ÷ 7.7 =	9.292	7.092	9.992	8.292
y	254.79 + 89.023 =	355.453	343.813	320.713	363.813
z	51.6 + 102.83 + 49.17 =	198.6	210.6	203.6	195.6

Add decimals

Remember

When adding and subtracting decimals, keep the decimal points under one another.

1 Copy and complete the following grid.

+	24.8	7.975	47.68	321.7	86.529	6.554	183.65	35.97	277.6
71.263									
8.575									
164.92									
37.96									
64.03									
8.074									

2 Add each group of decimal numbers.

- **a** 56.189 + 4.764 + 218.3 + 17.553 + 76.05 + 160.82
- **b** 34.088 + 116.4 + 27.65 + 44.7 + 126.09 + 28.661
- **c** 8.774 + 103.54 + 29.89 + 65.7 + 211.036 + 12.435
- **d** 135.06 + 74.1 + 67.015 + 4.877 + 166.9 + 39.682
- **e** 208.6 + 95.478 + 114.53 + 7.819 + 58.44 + 15.296
- **f** 4.837 + 245.8 + 79.72 + 39.053 + 164.7 + 81.089

3 What might the missing digits be in these calculations?

- **a** 6□.37□ + 41.2□3 = □□6.627
- **b** □.81□ + 27.□6 = □2.377
- **c** □6.□8 + 79.86□ = 13□.944
- **d** □.905 + 87.6□ = □2.□45
- **e** □9.□□3 + 58.766 = 9□.91□
- **f** 5□.96 + 72.8□ = 1□8.□1
- **g** 43.1□8 + □.57 = □1.72□
- **h** 9.□78 + □7.06□ = 10□.9□4
- **i** 8.6□4 + □3.□86 = 7□.56□
- **j** 2□□.8 + 386.□9 = □24.4□
- **k** 6□.5□ + □8.□5□ = 86.534
- **l** □18.□7 + 8□.6□5 = 3□6.70□

4 When you have worked out all the answers, write the letter from each box that has an answer between 180 and 210, then unjumble the letters to make the name of a place in Papua New Guinea.

a **P**	**b** **A**	**c** **B**	**d** **G**	**e** **D**
27.4 5.68 + 35.07 ______	171.06 + 35.8 ______	61.25 17.6 + 23.185 ______	66.237 + 138.56 ______	80.9 + 38.97 ______
f **T**	**g** **E**	**h** **S**	**i** **C**	**j** **V**
15.634 + 8.326 ______	176.22 + 30.6 ______	97.42 + 31.684 ______	97.8 5.96 + 23.67 ______	95.258 + 86.332 ______
k **O**	**l** **F**	**m** **N**	**n** **R**	**o** **K**
46.85 + 59.389 ______	38.625 + 29.351 ______	129.16 + 59.8 ______	4.85 216.7 + 39.67 ______	81.23 + 99.587 ______
p **W**	**q** **U**	**r** **Y**	**s** **I**	**t** **H**
52.672 + 23.811 ______	55.06 42.3 + 71.88 ______	78.268 + 19.83 ______	34.68 + 150.7 ______	27.51 + 78.656 ______

What is the place name?

Subtract decimals

1 Start from the number on the left and complete each column by subtracting the number shown at the top from the previous answer. The first one is done for you.

	Start	Subtract 12.85	Subtract 24.7	Subtract 8.549
a	97.04	84.19	59.49	50.941
b	116.5			
c	104.76			
d	83.261			
e	134.9			
f	232.18			
g	125.27			
h	78.063			
i	94.7			
j	228.59			
k	81.475			
l	148.6			

2 Subtract at each stage until you reach the number at the bottom. Did you get the same answers?

a
345.89
− 26.7
− 54.09
− 117.2
− 22.372
− 43.87
− 9.77
− 38.096
33.792

b
712.5
− 133.78
− 31.9
− 78.532
− 157.04
− 81.66
− 47.08
− 106.3
76.208

c
596.37
− 78.817
− 114.3
− 92.21
− 31.019
− 108.44
− 37.25
− 85.644
48.69

d
905.8
− 176.55
− 58.116
− 83.29
− 128.87
− 74.043
− 168.4
− 117.68
98.851

Subtract decimals

3 When you have worked out all the answers, write the letter from each box that has an answer between 75 and 90, then unjumble the letters to make the name of a place in Papua New Guinea.

a U	b D	c O	d G	e F
156.85 − 78.67 = ____	128.6 − 67.38 = ____	94.31 − 39.65 = ____	231.48 − 174.7 = ____	133.2 − 65.48 = ____
f C	**g A**	**h P**	**i K**	**j R**
81.553 − 36.818 = ____	143.05 − 56.82 = ____	281.3 − 126.89 = ____	85.326 − 47.09 = ____	191.55 − 114.93 = ____
k M	**l H**	**m L**	**n Y**	**o S**
320.4 − 173.15 = ____	203.117 − 85.26 = ____	182.91 − 103.59 = ____	114.81 − 78.932 = ____	105.06 − 37.2 = ____
p B	**q N**	**r T**	**s A**	**t W**
316.03 − 227.8 = ____	173.22 − 46.927 = ____	91.025 − 38.433 = ____	112.36 − 27.525 = ____	246.3 − 138.68 = ____

What is the place name?

Multiply decimals by 10, 100 and 1000

Remember

When multiplying by 10, the digits move one place to the left of the decimal point. When multiplying by 100, the digits move two places to the left of the decimal point. When multiplying by 1000, the digits move three places to the left of the decimal point.

1 Multiply the number on the left by 10, 100 and 1000.

	Number	× 10	× 100	× 1000
a	54.847			
b	2.559			
c	39.16			
d	78.925			
e	125.04			
f	62.58			
g	344.17			
h	63.9			
i	189.2			
j	21.625			
k	3.089			
l	257.4			
m	15.925			
n	242.07			
o	5.703			
p	7.68			
q	36.331			
r	29.476			

2 Multiply by 1000 to change kilometres to metres, multiply by 100 to change metres to centimetres, and multiply by 10 to change centimetres to millimetres.

a 27.813 km = _____ m
b 5.876 m = _____ cm
c 187.5 cm = _____ mm
d 54.76 m = _____ cm
e 28.75 cm = _____ mm
f 12.472 km = _____ m
g 246.87 cm = _____ mm
h 95.006 km = _____ m
i 140.35 m = _____ cm
j 84.03 km = _____ m
k 33.75 m = _____ cm
l 49.72 cm = _____ mm
m 216.182 m = _____ cm
n 109.61 cm = _____ mm
o 257.18 km = _____ m
p 9.816 cm = _____ mm
q 7.4 km = _____ m
r 565.445 m = _____ cm
s 105.8 km = _____ m
t 14.807 m = _____ cm
u 74.19 cm = _____ mm
v 50.452 m = _____ cm
w 110.8 cm = _____ mm
x 99.51 km = _____ m

Multiply decimals by multiples of 10

Help Box

When multiplying decimals by multiples of 10, first multiply by 10 and then multiply by the number of tens.

Example 1: 17.85 × 30

Calculate 17.85 × 10 = 178.5
Then multiply by 3: 178.5 × 3
The answer is 535.5

Example 2: 167.47 × 70

Calculate 167.47 × 10 = 1674.7
Then multiply by 7: 1674.7 × 7
The answer is 11 722.9

Work out the answer to each multiplication problem.

1 45.18 × 40 = ______

2 172.96 × 20 = ______

3 93.8 × 70 = ______

4 231.64 × 30 = ______

5 108.7 × 60 = ______

6 275.5 × 50 = ______

7 53.44 × 90 = ______

8 161.32 × 40 = ______

9 36.085 × 80 = ______

10 9.663 × 70 = ______

11 28.056 × 20 = ______

12 5.878 × 50 = ______

13 312.4 × 30 = ______

14 56.07 × 60 = ______

15 124.27 × 80 = ______

16 79.154 × 40 = ______

17 82.29 × 90 = ______

18 6.419 × 20 = ______

19 232.8 × 50 = ______

20 27.608 × 70 = ______

21 320.48 × 60 = ______

22 63.052 × 30 = ______

23 167.4 × 40 = ______

24 84.263 × 80 = ______

25 203.6 × 20 = ______

26 72.48 × 50 = ______

27 8.671 × 90 = ______

28 34.27 × 60 = ______

Multiply decimals by whole numbers

Help Box

When multiplying decimals by whole numbers other than 10 or multiples of 10, first estimate the answer so you know where the decimal point will go.

Example: 27.43 × 14

Estimate by rounding to 30 × 14 = 420
This indicates that the answer will be about but less than 420.

Remove the decimal point and calculate:
2743 × 14 = 38 402

Put the decimal point in the answer: 384.02

So, 27.43 × 14 = 384.02

Work out the answer to each multiplication problem, then add the five largest answers together.

1 39.4 × 6 = ____	**2** 8.97 × 9 = ____	**3** 17.85 × 5 = ____	**4** 63.7 × 3 = ____
5 81.06 × 4 = ____	**6** 217.6 × 8 = ____	**7** 9.378 × 7 = ____	**8** 6.593 × 5 = ____
9 46.7 × 18 = ____	**10** 51.9 × 14 = ____	**11** 8.68 × 19 = ____	**12** 54.5 × 17 = ____
13 39.65 × 15 = ____	**14** 93.86 × 25 = ____	**15** 8.463 × 23 = ____	**16** 6.892 × 16 = ____
17 26.23 × 34 = ____	**18** 8.747 × 29 = ____	**19** 113.8 × 37 = ____	
20 17.84 × 38 = ____	**21** 33.02 × 26 = ____	**22** 260.7 × 24 = ____	

1 ____
2 ____
3 ____
4 ____
5 ____

[ ]

Multiply decimals by decimals

Help Box

When multiplying decimals by decimals, first estimate the answer so you know where the decimal point will go.

Example: 37.85 × 9.3

Estimate by rounding to 38 × 9 = 342
This indicates that the answer will be about 342.

Remove the decimal point and calculate:
3785 × 93 = 352 005

Put the decimal point in the answer: 352.005

So, 37.85 × 9.3 = 352.005

1 Use rounding to decide where to place the decimal point in these multiplications.

a	23.7 × 7.8 = 18486	**b**	44.2 × 5.3 = 23426	**c**	38.12 × 6.97 = 2656964
d	14.55 × 11.8 = 17169	**e**	51.2 × 12.6 = 64512	**f**	52.75 × 9.81 = 5174775
g	25.6 × 3.9 = 9984	**h**	105.8 × 4.12 = 435896	**i**	45.12 × 3.9 = 175968
j	19.86 × 5.03 = 99896	**k**	23.4 × 8.02 = 187668	**l**	63.7 × 5.4 = 34398

2 How many balloons can you burst?

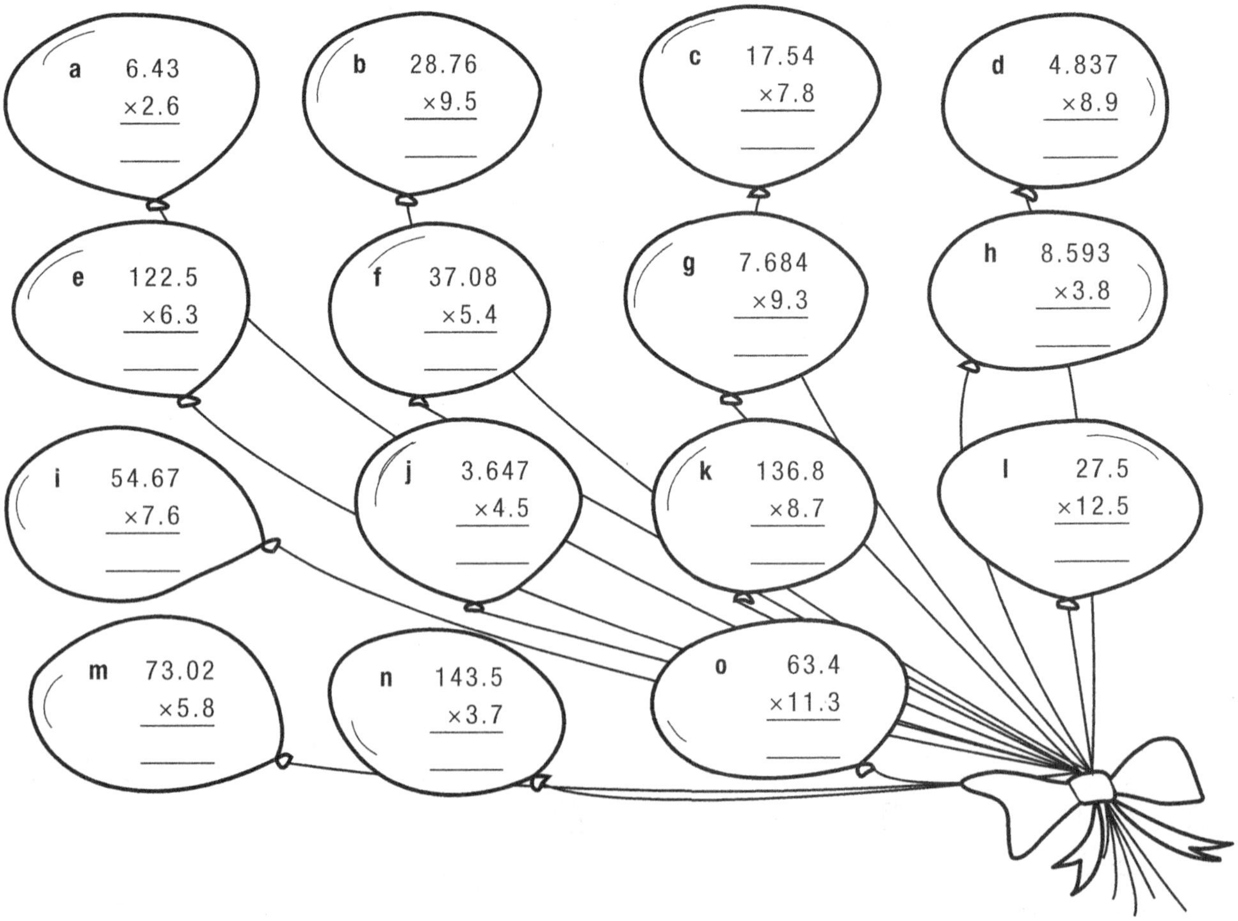

a 6.43 × 2.6 = ____

b 28.76 × 9.5 = ____

c 17.54 × 7.8 = ____

d 4.837 × 8.9 = ____

e 122.5 × 6.3 = ____

f 37.08 × 5.4 = ____

g 7.684 × 9.3 = ____

h 8.593 × 3.8 = ____

i 54.67 × 7.6 = ____

j 3.647 × 4.5 = ____

k 136.8 × 8.7 = ____

l 27.5 × 12.5 = ____

m 73.02 × 5.8 = ____

n 143.5 × 3.7 = ____

o 63.4 × 11.3 = ____

Divide decimals by 10, 100 and 1000

Remember

When dividing by 10, the digits move one place to the right of the decimal point. When dividing by 100, the digits move two places to the right of the decimal point. When dividing by 1000, the digits move three places to the right of the decimal point.

1 Divide the number on the left by 10, 100 and 1000.

	Number	÷ 10	÷ 100	÷ 1000
a	718.6			
b	9672.64			
c	438.427			
d	3257.8			
e	316.08			
f	206.25			
g	1534.66			
h	410.9			
i	765.28			
j	374.5			
k	809.54			
l	7629.4			
m	22.093			
n	54.13			
o	688.2			
p	17.551			
q	253.917			
r	38.4			

2 Divide by 1000 to change metres to kilometres, divide by 100 to change centimetres to metres, and divide by 10 to change millimetres to centimetres.

a 45670 m = _____ km
b 1165 cm = _____ m
c 254 mm = _____ cm
d 382.7 cm = _____ m
e 63.25 mm = _____ cm
f 2375.4 m = _____ km
g 14.49 mm = _____ cm
h 885.12 m = _____ km
i 648.2 cm = _____ m
j 3803.6 m = _____ km
k 409.1 cm = _____ m
l 178.5 mm = _____ cm
m 57.08 cm = _____ m
n 77.6 mm = _____ cm
o 6328 m = _____ km
p 8.71 mm = _____ cm
q 8209.8 m = _____ km
r 968 cm = _____ m
s 17809 m = _____ km
t 5.74 cm = _____ m
u 9132.1 mm = _____ cm
v 8.5 cm = _____ m
w 1506 mm = _____ cm
x 3359.22 m = _____ km

Divide decimals by multiples of 10

Help Box

When dividing decimals by multiples of 10, first divide by 10 and then divide by the number of tens.

Example 1: 942.4 ÷ 40

Calculate 942.4 ÷ 10 = 94.24
Then divide by 4: 94.24 ÷ 4
The answer is 23.56

Example 2: 1837.5 ÷ 50

Calculate 1837.5 ÷ 10 = 183.75
Then divide by 7: 183.75 ÷ 5
The answer is 36.75

When you have calculated all the answers, find the total of each column and add them together. Do they equal the GRAND TOTAL?

1 20)56.34	**2** 40)452.8	**3** 70)610.4	**4** 30)771.9	**5** 50)608.5
6 30)131.16	**7** 20)122.8	**8** 60)21.36	**9** 40)631.2	**10** 30)117.9
11 50)90.2	**12** 80)365.6	**13** 20)654.6	**14** 70)216.09	**15** 40)531.6
16 30)64.65	**17** 20)837.6	**18** 60)1621.8	**19** 40)658.8	**20** 50)1423.5

GRAND TOTAL

262.821

Divide decimals by whole numbers

Help Box

Example 1: 365.04 ÷ 8

$$\begin{array}{r} 45.63 \\ 8\overline{)3\,6^{4}5\,.^{5}0^{2}4} \end{array}$$

When calculating it is important to place the decimal point at the place above where it appears during the division.

Example 2: 467.55 ÷ 15

$$\begin{array}{r} 31.17 \\ 15\overline{)467.55} \\ -45 \\ \hline 17 \\ -15 \\ \hline 25 \\ -15 \\ \hline 105 \\ -105 \\ \hline 0 \end{array}$$

Calculate the answers to find the lucky ticket.

LUCKY TICKET
27.8

LUCKY TICKET 27.8	**1** $5\overline{)65.75}$	**2** $7\overline{)937.3}$	**3** $3\overline{)851.4}$
4 $8\overline{)172.64}$	**5** $4\overline{)56.76}$	**6** $6\overline{)878.16}$	**7** $9\overline{)552.78}$
8 $12\overline{)853.92}$	**9** $15\overline{)558.75}$	**10** $21\overline{)187.74}$	**11** $17\overline{)753.1}$
12 $24\overline{)289.92}$	**13** $14\overline{)663.46}$	**14** $32\overline{)889.6}$	**15** $18\overline{)804.6}$
16 $25\overline{)903.5}$	**17** $19\overline{)830.49}$	**18** $23\overline{)410.78}$	**19** $36\overline{)426.6}$
20 $16\overline{)731.2}$	**21** $22\overline{)682.88}$	**22** $35\overline{)784.35}$	**23** $27\overline{)676.08}$

Divide decimals by decimals

Help Box

When dividing decimals by decimals, first estimate the answer so you know where the decimal point will go.

Example: 15.614 ÷ 3.7

Estimate by rounding to 16 ÷ 4 = 4

This indicates that the answer will be about 4.

Remove the decimal point and calculate:
15 614 ÷ 37 = 422

Put the decimal point in the answer: 4.22

So, 15.614 ÷ 3.7 = 4.22

How many goals can you shoot? Calculate each answer to two decimal places.

1 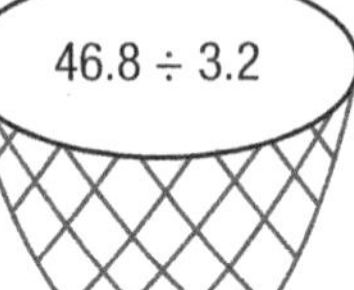

2 

3 

4 

5 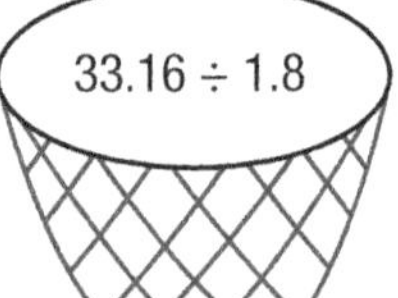

6 

7 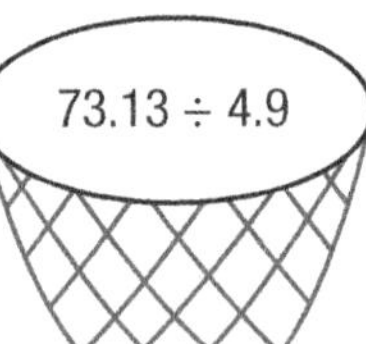

8 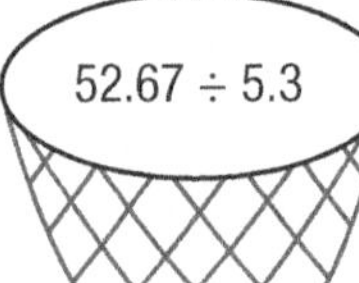

9 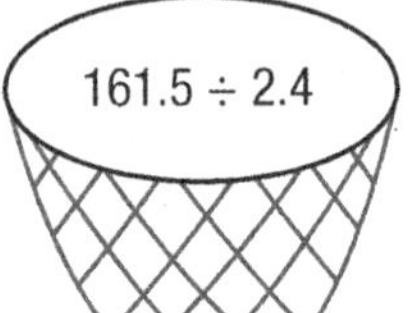

10 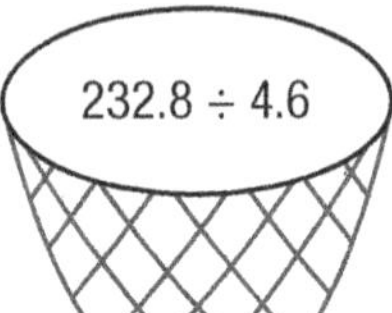

11 

12 

13 

14 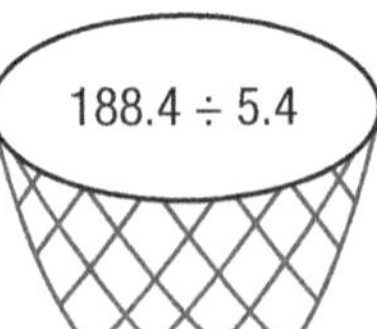

15 

16 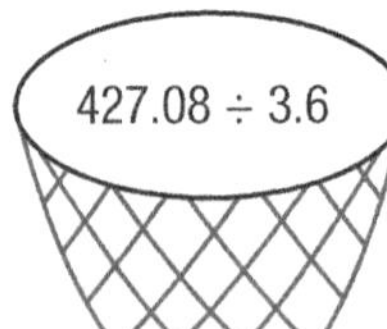

17 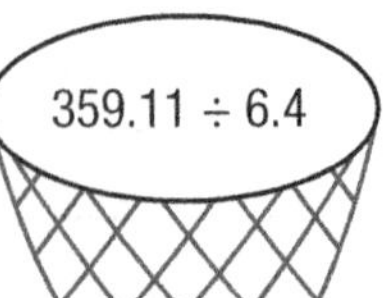

18 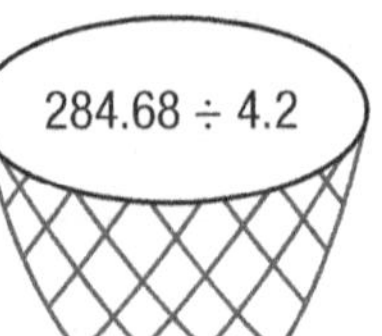

19 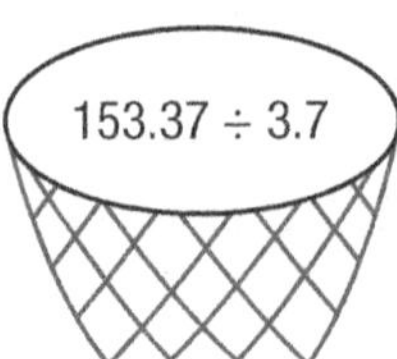

20 

Use a calculator for division

Play this game with a friend. You will need a calculator. Each player needs a set of counters (each set a different colour).

Take turns to choose two numbers from the cloud and use a calculator to divide the numbers. If the answer is on the grid, place a counter on it. Only one counter may be placed on each number.

The winner is the first player to make a continuous line of counters from one side of the grid to the other.

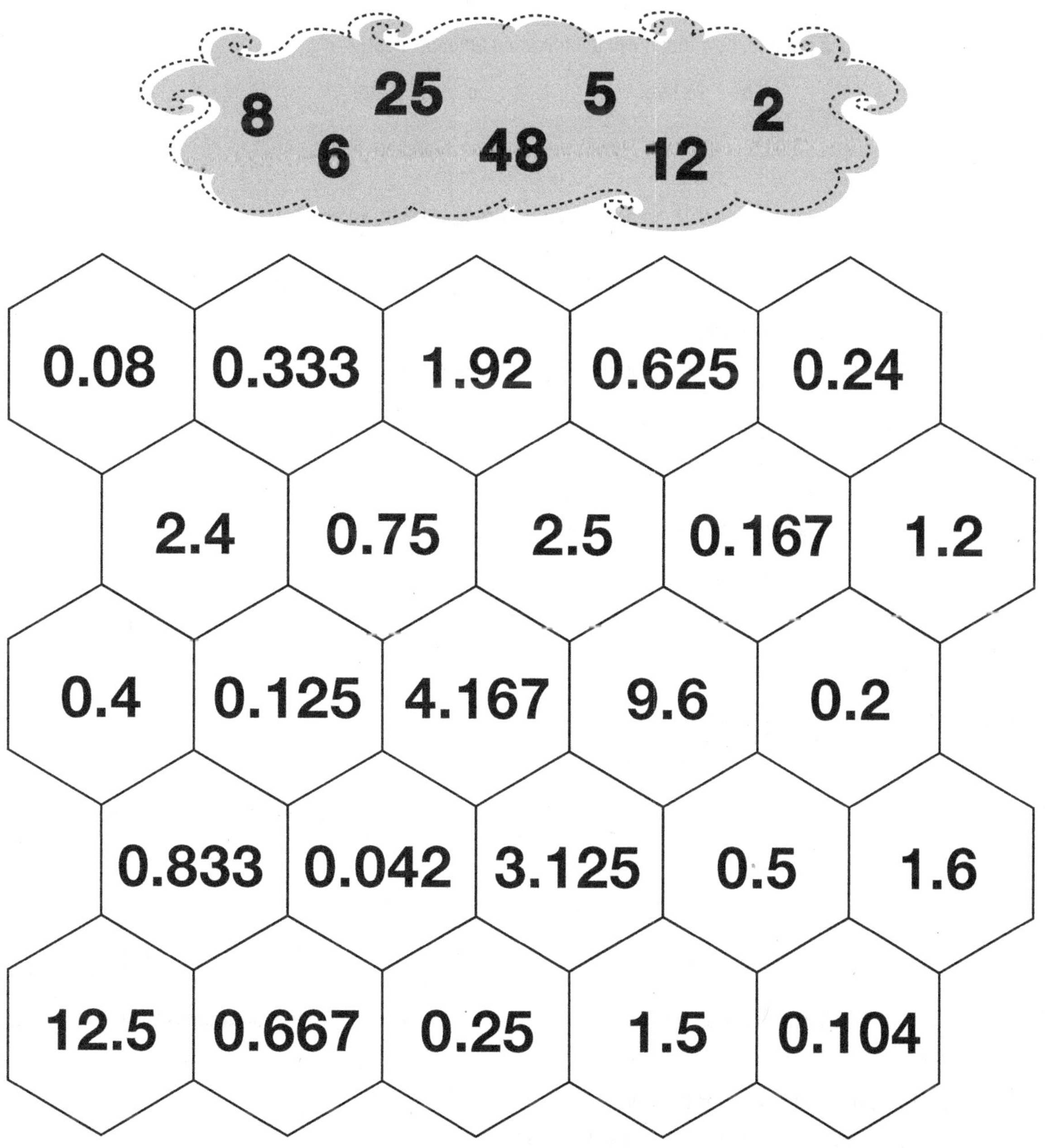

Solve decimal word problems

Decide what process to use and then solve each word problem. Where appropriate, round answers to two decimal places.

1 Six nights accommodation costs K921.30.

- **a** What is the cost per night?
- **b** At the same rate per night, how much would it cost to stay for 12 nights?

2 If petrol costs 73.8 toea per litre, how much will the following amounts cost?

a 45 litres **b** 28 litres **c** 53.2 litres **d** 37.7 litres

3 Fourteen people share a K893.90 lottery win. How much does each person get?

4 How much would 2.3 kg of beef at K9.75 kg and 5.4 kg of carrots at K3.95 kg cost?

5 A car has four passengers weighing 76.7 kg, 103.25 kg, 49.86 kg and 88.39 kg respectively.

- **a** What is the total weight of passengers in the car?
- **b** How much heavier is the heaviest person than the lightest person?

6 Mona had K834.19 in her bank account. If she withdraws K267.65 to pay for repairs to her house, how much will be left in her account?

7 At a market, a gardener sold 47 trees at K17.35 each.

- **a** How much money did he receive?
- **b** If it cost him K258.60 to prepare the trees for sale, how much profit did he make?

8 A car travelled 473.9 km in seven hours.

- **a** What was its average speed per hour?
- **b** At the same speed, how far would it travel in nine hours?

9 A family has to travel a total distance of 816.42 km. If they have already travelled 568.7 km, how much further do they have left to travel?

10 Each box of books weighs 15.66 kg. What is the combined weight of:

a 30 boxes? **b** 50 boxes? **c** 18 boxes? **d** 44 boxes?

11 A village used 877.5 litres of water on Monday, 709.35 litres on Tuesday, 963.27 litres on Wednesday and 782.8 litres on Thursday.

- **a** How much water was used over the four days?
- **b** What was the average amount of water used per day?

7.1.3 Convert between fractions, decimals and percentages

Convert between fractions and decimals

Help Box

- Converting fractions with denominators of 10, 100 or 1000 to decimals

  Examples: $\frac{6}{10} = 0.6$, $\frac{37}{100} = 0.37$, $\frac{215}{1000} = 0.215$
- Converting fractions with denominators that are easy to change to 10, 100 or 1000 to decimals

  Examples: $\frac{3}{5} = \frac{6}{10} = 0.6$, $\frac{7}{50} = \frac{14}{100} = 0.14$, $\frac{5}{8} = \frac{625}{1000} = 0.625$
- Converting fractions with denominators that cannot be changed to 10, 100 or 1000 to decimals

  Examples: $\frac{2}{7}$, $\frac{5}{6}$, $\frac{7}{12}$

  To do these, divide the numerator by the denominator:

  $\frac{2}{7} = 2 \div 7 = 0.29$ (rounded to two decimal places)

1 Change these fractions with denominators of 10, 100 or 1000 to decimals.

a	$\frac{54}{100}$	b	$\frac{7}{10}$	c	$\frac{31}{100}$	d	$\frac{325}{1000}$	e	$\frac{503}{1000}$	f	$\frac{63}{100}$
g	$\frac{9}{10}$	h	$\frac{437}{1000}$	i	$\frac{4}{100}$	j	$\frac{67}{1000}$	k	$\frac{194}{1000}$	l	$\frac{14}{1000}$
m	$\frac{80}{100}$	n	$\frac{40}{1000}$	o	$\frac{829}{1000}$	p	$\frac{5}{100}$	q	$\frac{13}{100}$	r	$\frac{346}{1000}$
s	$\frac{7}{1000}$	t	$\frac{45}{100}$	u	$\frac{87}{1000}$	v	$\frac{9}{1000}$	w	$\frac{1}{100}$	x	$\frac{1}{1000}$

2 Change these fractions with denominators that can be changed to 10, 100 or 1000 to decimals.

a	$\frac{4}{5}$	b	$\frac{13}{20}$	c	$\frac{5}{8}$	d	$\frac{3}{4}$	e	$\frac{2}{5}$	f	$\frac{12}{50}$
g	$\frac{1}{8}$	h	$\frac{7}{25}$	i	$\frac{18}{30}$	j	$\frac{36}{60}$	k	$\frac{1}{20}$	l	$\frac{3}{8}$
m	$\frac{29}{50}$	n	$\frac{9}{20}$	o	$\frac{19}{25}$	p	$\frac{8}{40}$	q	$\frac{278}{500}$	r	$\frac{17}{125}$
s	$\frac{81}{90}$	t	$\frac{134}{500}$	u	$\frac{51}{125}$	v	$\frac{7}{8}$	w	$\frac{22}{25}$	x	$\frac{9}{30}$

3 Divide the numerator by the denominator to change these fractions to decimals, and round each answer to two decimal places.

a	$\frac{5}{6}$	b	$\frac{7}{12}$	c	$\frac{4}{9}$	d	$\frac{3}{7}$	e	$\frac{7}{9}$	f	$\frac{8}{11}$
g	$\frac{1}{3}$	h	$\frac{4}{11}$	i	$\frac{2}{15}$	j	$\frac{5}{9}$	k	$\frac{2}{11}$	l	$\frac{7}{15}$
m	$\frac{5}{7}$	n	$\frac{1}{6}$	o	$\frac{2}{9}$	p	$\frac{7}{11}$	q	$\frac{11}{12}$	r	$\frac{4}{7}$
s	$\frac{1}{9}$	t	$\frac{5}{12}$	u	$\frac{1}{11}$	v	$\frac{6}{7}$	w	$\frac{8}{9}$	x	$\frac{2}{3}$

4 Write these mixed numbers and improper fractions as decimals.

a	$\frac{23}{10}$	b	$\frac{153}{100}$	c	$2\frac{7}{1000}$	d	$1\frac{85}{100}$	e	$\frac{39}{10}$	f	$\frac{539}{100}$
g	$5\frac{8}{100}$	h	$\frac{3244}{1000}$	i	$7\frac{61}{100}$	j	$3\frac{95}{1000}$	k	$\frac{8}{5}$	l	$1\frac{3}{8}$
m	$6\frac{7}{8}$	n	$3\frac{12}{25}$	o	$5\frac{9}{20}$	p	$\frac{19}{4}$	q	$\frac{15}{8}$	r	$\frac{37}{5}$
s	$\frac{61}{20}$	t	$\frac{167}{25}$	u	$2\frac{21}{25}$	v	$\frac{27}{8}$	w	$\frac{43}{4}$	x	$\frac{121}{10}$

5 Write these decimals as fractions in their simplest form.

a	0.8	b	0.513	c	0.71	d	1.5	e	3.12	f	0.243
g	0.023	h	5.07	i	2.15	j	1.009	k	6.2	l	8.22
m	3.857	n	7.93	o	4.6	p	5.088	q	1.275	r	0.155
s	2.65	t	4.324	u	9.016	v	1.85	w	3.006	x	2.418

Convert between fractions, decimals and percentages

Remember

Fractions with a denominator of 100 can be changed directly to percentages.

Examples: $\frac{22}{100} = 22\%$; $1\frac{34}{100} = 134\%$; $0.17 = \frac{17}{100} = 17\%$; $2.76 = 2\frac{76}{100} = 276\%$

Fractions with denominators such as 5, 10, 20, 25 and 50 can be changed to percentages by making an equivalent fraction out of 100.

Examples: $\frac{4}{5} = \frac{80}{100} = 80\%$; $\frac{11}{25} = \frac{44}{100} = 44\%$

1 Change these fractions and decimals to percentages.

a	$\frac{27}{100}$	b	$\frac{63}{100}$	c	0.11	d	0.53	e	0.68	f	$\frac{41}{100}$
g	1.55	h	$\frac{7}{100}$	i	2.08	j	$1\frac{65}{100}$	k	$2\frac{33}{100}$	l	0.83
m	2.19	n	1.05	o	$\frac{2}{5}$	p	$\frac{16}{50}$	q	$\frac{4}{25}$	r	$\frac{14}{20}$
s	$1\frac{4}{5}$	t	$2\frac{6}{25}$	u	$1\frac{7}{20}$	v	0.7	w	1.4	x	$\frac{31}{50}$

Help Box

When you change a fraction or decimal to a percentage it is the same as multiplying by 100.

Examples:

1. $\frac{21}{100} \times \frac{100}{1} = \frac{2100}{100} = 21\%$ *or* $0.21 \times 100 = 21\%$
2. $2\frac{3}{5} = \frac{13}{5} \times \frac{100}{1} = \frac{1300}{5} = 260\%$ *or* $2.6 \times 100 = 260\%$
3. $\frac{3}{8} \times \frac{100}{1} = \frac{300}{8} = 37.5\%$ *or* $37\frac{1}{2}\%$

2 Multiply by 100 to change these fractions and decimals to percentages.

a	0.9	**b**	1.31	**c**	$\frac{1}{4}$	**d**	$\frac{4}{5}$	**e**	$\frac{3}{20}$	**f**	2.08
g	$1\frac{3}{4}$	**h**	1.5	**i**	6.73	**j**	$2\frac{21}{25}$	**k**	0.007	**l**	1.2
m	$\frac{47}{50}$	**n**	$\frac{6}{20}$	**o**	$\frac{14}{25}$	**p**	$\frac{5}{8}$	**q**	$\frac{2}{5}$	**r**	$\frac{39}{50}$
s	1.61	**t**	$\frac{7}{25}$	**u**	5.035	**v**	$4\frac{3}{5}$	**w**	$1\frac{7}{10}$	**x**	0.01

Help Box

When you change a percentage to a decimal or a fraction it is the same as dividing by 100.

Examples – decimals:

1. $8\% = 8 \div 100 = 0.08$
2. $20\% = 20 \div 100 = 0.2$
3. $26\frac{1}{2}\% = 26.5 \div 100 = 0.265$

Examples – fractions:

1. $8\% = \frac{8}{100} = \frac{2}{25}$
2. $20\% = \frac{20}{100} = \frac{1}{5}$
3. $26\frac{1}{2}\% = \frac{53}{200}$

3 Divide by 100 to change these percentages to decimals.

a	57%	**b**	32%	**c**	73%	**d**	14%	**e**	7%	**f**	94%
g	81%	**h**	163%	**i**	4%	**j**	217%	**k**	28%	**l**	135%
m	$54\frac{1}{2}\%$	**n**	40%	**o**	$24\frac{1}{2}\%$	**p**	$17\frac{1}{4}\%$	**q**	$39\frac{3}{4}\%$	**r**	86%
s	1%	**t**	100%	**u**	209%	**v**	146%	**w**	90%	**x**	$33\frac{1}{4}\%$

4 Divide by 100 to change these percentages to fractions, and write each fraction in its simplest form.

a	10%	**b**	40%	**c**	6%	**d**	70%	**e**	15%	**f**	48%
g	64%	**h**	26%	**i**	72%	**j**	38%	**k**	95%	**l**	120%
m	250%	**n**	175%	**o**	14%	**p**	66%	**q**	4%	**r**	86%
s	$33\frac{1}{2}\%$	**t**	$17\frac{1}{2}\%$	**u**	$45\frac{1}{2}\%$	**v**	$62\frac{1}{2}\%$	**w**	$11\frac{1}{4}\%$	**x**	$36\frac{1}{4}\%$

5 Copy and complete these charts. Write each fraction in its simplest form.

a

Fraction	Decimal	Percentage
$\frac{4}{5}$		
		40%
	0.09	
		25%
$\frac{7}{8}$		
	0.85	
$\frac{9}{20}$		
		$12\frac{1}{2}\%$
$\frac{7}{25}$		
		70%
	1.07	
	0.005	

b

Fraction	Decimal	Percentage
$\frac{3}{4}$		
	0.56	
		$37\frac{1}{2}\%$
	0.03	
$\frac{15}{50}$		
		1%
	1.4	
$1\frac{3}{5}$		
		171%
		50%
	0.008	
$2\frac{3}{8}$		

6 Write the fraction or percentage that is equivalent to the fraction or percentage on the left.

a 0.6 $\frac{6}{100}$ $\frac{6}{10}$ $\frac{3}{4}$
b $\frac{1}{3}$ $33\frac{1}{3}\%$ 0.3 0.13
c $\frac{3}{4}$ 0.34 75% $\frac{6}{9}$
d 27% $\frac{2}{7}$ 0.2 $\frac{27}{100}$
e 50% $\frac{5}{100}$ $\frac{1}{2}$ 0.05
f $\frac{7}{10}$ 0.7 7% $\frac{1}{7}$
g $\frac{1}{4}$ 0.14 $\frac{4}{10}$ 25%
h 0.2 2% $\frac{20}{100}$ $\frac{1}{2}$
i $\frac{6}{8}$ $\frac{3}{4}$ 0.6 65%
j 10% $\frac{10}{10}$ $\frac{1}{10}$ 0.01
k $\frac{1}{5}$ 5% 0.2 $\frac{5}{10}$
l $\frac{3}{10}$ $\frac{6}{10}$ 0.31 30%
m $\frac{2}{3}$ 0.6 $66\frac{2}{3}\%$ 0.23
n $62\frac{1}{2}\%$ 0.625 6.25 62.5
o 0.009 0.9 0.9% 9%

7 Write the fraction or percentage that is not equivalent to the others.

a 0.9 90% 0.09 $\frac{9}{10}$
b $\frac{2}{5}$ 40% $\frac{4}{10}$ 0.2
c 50% 0.05 5% $\frac{5}{100}$
d $\frac{3}{4}$ 70% 75% 0.75
e 0.6 $\frac{3}{5}$ $\frac{6}{10}$ 6%
f 24% $\frac{2}{4}$ $\frac{24}{100}$ 0.24
g $\frac{6}{12}$ 0.5 50% 0.6
h 0.85 $\frac{8}{10}$ 0.8 80%
i 25% 0.14 $\frac{1}{4}$ 0.25
j $\frac{60}{100}$ 6% 0.6 $\frac{6}{10}$
k 7% 0.07 $\frac{7}{10}$ $\frac{7}{100}$
l $\frac{1}{10}$ 1% 10% $\frac{10}{100}$
m 0.35 35% $\frac{7}{20}$ $3\frac{1}{2}\%$
n $42\frac{1}{2}\%$ 42.5% 42.5 0.425
o 0.017 1.7% 17% $\frac{17}{1000}$

7.1.4 Use percentage in a variety of real life situations

Calculate discount and selling price

Help Box

There are two ways to calculate discount:

1. Change the percent discount to a fraction and then multiply:
   10% of K70 = $\frac{1}{10} \times \frac{70}{1} = \frac{70}{10}$ = K7

2. Change the percent discount to a decimal and then multiply:
   10% of K70 = 0.1 × 70 = 7.0 = K7

Copy and complete these tables by using one of the above methods.

1

	Full price	Percentage discount	Discount	Selling price
a	K90	10%		
b	K55	20%		
c	K25	5%		
d	K150	30%		
e	K48	25%		
f	K100	$12\frac{1}{2}$%		
g	K80	5%		
h	K100	$7\frac{1}{2}$%		
i	K175	25%		
j	K120	15%		
k	K37.50	10%		
l	K62.50	20%		
m	K240	$12\frac{1}{2}$%		
n	K96	15%		
o	K165	5%		
p	K140	$7\frac{1}{2}$%		
q	K23.50	10%		
r	K85	25%		
s	K110	15%		

2

	Full price	Percentage discount	Discount	Selling price
a	K300	25%		
b	K260	15%		
c	K175	5%		
d	K215	10%		
e	K285	8%		
f	K110	4%		
g	K330	$17\frac{1}{2}$%		
h	K270	15%		
i	K405	20%		
j	K390	$12\frac{1}{2}$%		
k	K220	6%		
l	K112	$7\frac{1}{2}$%		
m	K345	15%		
n	K310	35%		
o	K275	25%		
p	K160	$17\frac{1}{2}$%		
q	K236	20%		
r	K187.50	10%		
s	K124.50	30%		

Calculate percentage increase

Example 1: To increase K12 by 10%
First find 10% of K12: $0.1 \times 12 = 1.20$,
then add K1.20 to K12 = K13.20
So, K12 increased by 10% is K13.20

Example 2: To increase 40 by 25%
First find 25% of 40: $0.25 \times 40 = 10$,
then add 10 to 40 = 50
So, 40 increased by 25% is 50

1 Copy and complete this table to show the difference between cost price and sale price.

	Cost price	Percentage increase	Amount of increase	Sale price
a	K50	15%		
b	K85	25%		
c	K63	10%		
d	K144	$12\frac{1}{2}\%$		
e	K550	30%		
f	K375	20%		
g	K1560	15%		
h	K2850	5%		
i	K1076	$12\frac{1}{2}\%$		
j	K2330	25%		

2 Increase each of the following amounts by 15%.

a 180 g b 150 g c 210 g d 500 g e 360 g f 420 g
g 730 g h 1100 g i 2600 g j 1950 g k 3250 g l 2250 g
m 1870 g n 2050 g o 1790 g p 3080 g q 2950 g r 1560 g

3 Increase each of the following quantities by 20%.

a 840 mL b 615 mL c 1010 mL d 725 mL e 1250 mL f 675 mL
g 2340 mL h 1835 mL i 2990 mL j 3125 mL k 2770 mL l 805 mL
m 995 mL n 2010 mL o 3675 mL p 4005 mL q 2285 mL r 915 mL

4 Increase each of the following lengths by 5%.

a 150 cm b 340 cm c 270 cm d 770 cm e 560 cm f 410 cm
g 890 cm h 650 cm i 930 cm j 1240 cm k 1610 cm l 2160 cm
m 1080 cm n 3250 cm o 2330 cm p 1750 cm q 3890 cm r 2040 cm

5 Increase each of the following amounts by $7\frac{1}{2}\%$.

a	K56	**b**	K124	**c**	K92	**d**	K248	**e**	K188	**f**	K200
g	K340	**h**	K110	**i**	K246	**j**	K370	**k**	K504	**l**	K422
m	K1000	**n**	K2200	**o**	K1680	**p**	K2960	**q**	K1390	**r**	K3810

Solve percentage word problems

1 Lydia bought some new clothes at a sale. The total price was K133 less 15%. How much did Lydia pay for the clothes?

2 In an orchard of 460 trees, 20% are guava trees, 25% are mango trees and the rest are orange trees.

- **a** How many are guava trees?
- **b** How many are mango trees?
- **c** How many are orange trees?

3 At a rugby match, there were 4300 spectators. Children made up 10% of this number. How many adults were there?

4 A sales person made K4650. She was paid this amount plus 5% as a bonus or commission.

- **a** How much was her bonus?
- **b** How much was she paid in total?

5 In a village of 650 people, 84% have influenza. How many people are healthy?

6 Jack wants to increase the ingredients in a recipe by 25%. How much of each ingredient should he use?

- **a** 200 g flour
- **b** 140 g sugar
- **c** 70 g cocoa
- **d** 120 g butter

7 A company made a profit of K86 000. If it increases its profit by these percentages, how much will it make?

- **a** 20%
- **b** 5%
- **c** 15%
- **d** $12\frac{1}{2}\%$
- **e** 7%
- **f** 2%

8 The attendance at a school increased by 15% in the last year. If the attendance before the increase was 180 students, what is the attendance now?

9 A water tank that holds 12 500 litres is at 75% capacity.

- **a** How much water is in the tank?
- **b** How much more water will the tank take before it is full?

10 A car was advertised for sale at K16 700, but the seller reduced the price by 4% when the buyer offered cash. What did the buyer pay for the car?

11 At the start of the year, Naomi's height was 140 cm. By the end of the year, her height had increased by 5%. How tall was Naomi at the end of the year?

12 For one day only, a cinema was offering tickets at 30% discount. If the normal price of a ticket was K10.50, how much was a discounted ticket?

13 The manager of a company earned K33 600 for a year. He had to pay 27% of this in tax.

a How much tax did he pay? **b** What was his income after tax?

14 A refrigerator costing K890 had its price reduced by $12\frac{1}{2}$%. How much did the refrigerator cost?

15 A car salesman sold a car for K12 990 and was paid a commission of $5\frac{1}{2}$%. How much was his commission?

16 A sports stadium was enlarged so that its capacity was increased by 25%. If it held 24 500 spectators before it was enlarged, how many spectators will it hold now?

17 On a long trip, a car used 80% of its full tank of 70 litres of petrol. How much petrol remained in the tank?

18 A new car's price was increased by $7\frac{1}{2}$%. If its original price was K33 000, what was its new price?

19 A distance runner improved her best time of 150 minutes by 5%. What was her new best time?

20 A school took 15% of its students on a day trip. There are 340 students at the school.

a How many students went on the day trip? **b** How many students remained at school?

21 At the beginning of the year, a tree was 150 centimetres tall. At the end of the year, the tree's height had increased by $8\frac{1}{2}$%. What was its new height?

22 A hotel was sold for K225 000. The buyers had to pay a 20% deposit. How much deposit did they pay?

23 Charles bought a new motorbike for K4800. After one year, it lost 9% of its value.

a What is the value of the motorbike after one year?

b If it loses a further 9% after the second year, what will be its value?

24 In a school of 260 students, 45% are boys and the rest are girls.

a How many students are girls? **b** How many students are boys?

7.1.5 Convert between ratios and fractions

Write ratios using a sign or as a fraction

Help Box

Ratios can be written using a colon sign (:) or as a fraction.

Examples: A lemon drink can be made by mixing 1 cup of lemon juice with 5 cups of water. This can be written as 1:5 or $\frac{1}{5}$.

On a map, 1 cm represents 100 cm on the ground. This can be written as 1:100 or $\frac{1}{100}$.

1 Write each of these as a ratio and a fraction.

- **a** the number of dogs to the number of pigs
- **b** the number of pigs to the number of chickens
- **c** the number of chickens to the number of dogs
- **d** the number of dogs to the total number of animals
- **e** the number of chickens to the total number of animals
- **f** the number of pigs to the total number of animals

2 In a grade of 28 students, 15 walk to school, 6 catch a PMV and 7 ride bikes. Write each of these as a ratio and a fraction.

- **a** the number who walk to the number who catch a PMV
- **b** the number who ride bikes to the number who walk
- **c** the number who walk to the total number of students
- **d** the number who catch a PMV to the number who ride bikes
- **e** the number who ride bikes to the total number of students
- **f** the number who catch a PMV to the total number of students

3 In a box of shapes there were 8 triangles, 11 squares, 5 circles and 9 hexagons. Write each of these as a ratio and a fraction.

- **a** the number of hexagons to the number of circles
- **b** the number of triangles to the number of squares
- **c** the number of squares to the number of circles
- **d** the number of circles to the number of triangles
- **e** the number of squares to the total number of shapes
- **f** the number of triangles to the total number of shapes

4 Write each of these as ratios and as fractions.

a	1 mm to 7 mm	b	K10 to K3	c	8 mL to 17 mL	d	5 kg to 2 kg
e	5 cm to 50 cm	f	12 g to 27 g	g	30 t to 85 t	h	45 km to 8 km
i	8 L to 20 L	j	12 m to 5 m	k	K25 to K20	l	15 mm to 18 mm
m	24 cm to 16 cm	n	3 kg to 16 kg	o	14 mL to 20 mL	p	K30 to K45

5 Write each of these fractions as a ratio using the : sign.

a	$\frac{1}{3}$	b	$\frac{2}{7}$	c	$\frac{5}{3}$	d	$\frac{7}{10}$	e	$\frac{4}{1}$	f	$\frac{5}{6}$
g	$\frac{3}{7}$	h	$\frac{9}{8}$	i	$\frac{10}{3}$	j	$\frac{5}{9}$	k	$\frac{11}{6}$	l	$\frac{1}{8}$
m	$\frac{1}{2}$	n	$\frac{4}{3}$	o	$\frac{12}{7}$	p	$\frac{8}{11}$	q	$\frac{9}{15}$	r	$\frac{6}{1}$

6 Write each of these ratios as fractions.

a	6:11	b	4:5	c	7:4	d	13:15	e	5:7	f	8:5
g	3:9	h	12:5	i	1:5	j	10:9	k	8:17	l	11:4
m	5:8	n	9:20	o	6:7	p	5:12	q	19:8	r	15:21

Simplify ratios

Help Box

In the same way as fractions, ratios can sometimes be simplified or reduced to an equivalent ratio.

Example: A ratio of 2:8 can be simplified to 1:4 by dividing both numbers by 2. A ratio of 3:9 can be simplified to 1:3 by dividing both numbers by 3.

1 Write each of these ratios in their simplest form.

a	4:6	b	3:12	c	15:10	d	6:18	e	33:22	f	4:12
g	2:10	h	4:20	i	12:6	j	3:15	k	7:14	l	75:25
m	24:72	n	18:15	o	21:24	p	12:60	q	16:40	r	12:9
s	8:32	t	36:12	u	18:30	v	8:12	w	20:16	x	16:24

2 Write each of these as a ratio in its simplest form.

a	K3 to K12	**b**	500 g to 250 g	**c**	25 mL to 150 mL	**d**	60 cm to 80 cm
e	8 hours to 24 hours	**f**	24 kg to 64 kg	**g**	55t to 10t	**h**	75 mm to 30 mm
i	K21 to K14	**j**	14 L to 63 L	**k**	88 m to 33 m	**l**	10 days to 14 days
m	350 g to 600 g	**n**	40 mL to 16 mL	**o**	45 cm to 60 cm	**p**	16 hours to 12 hours
q	15t to 50t	**r**	15 L to 9 L				

3 Write the simplest ratio for the shaded parts : unshaded parts in these diagrams.

a 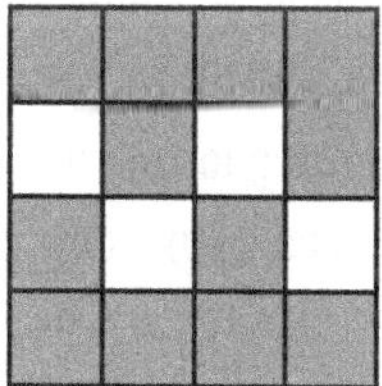

b 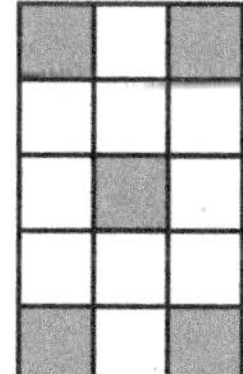

c 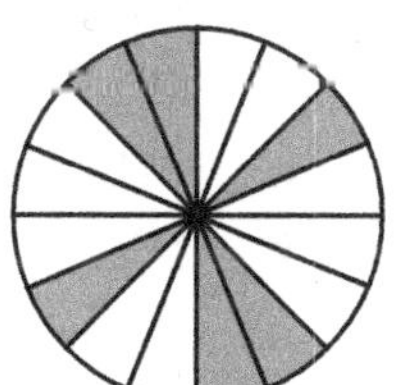

d 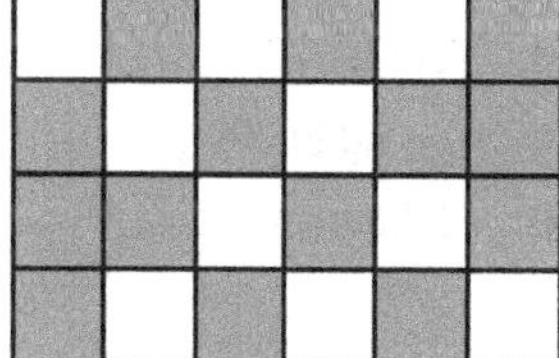

e 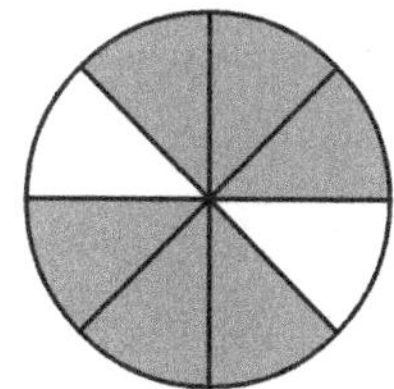

f 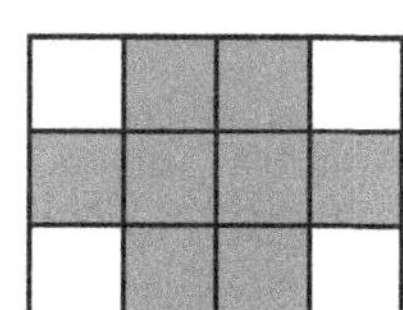

g 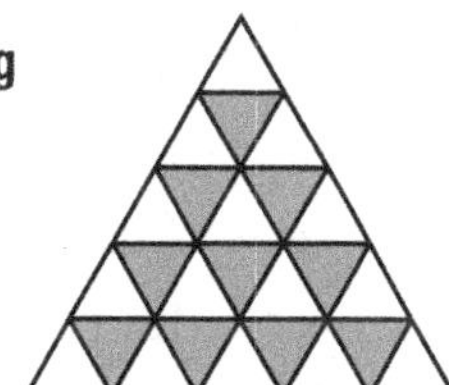

h 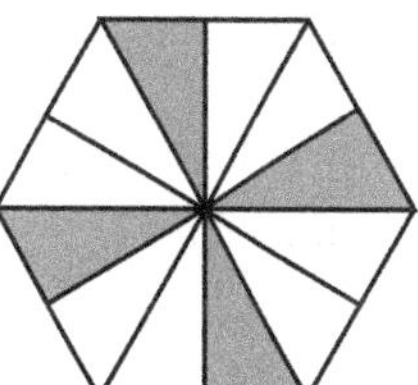

4 Write the simplest ratio for the shaded parts : whole shape in these diagrams.

a 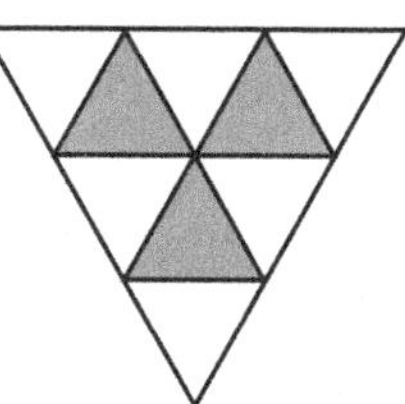

b 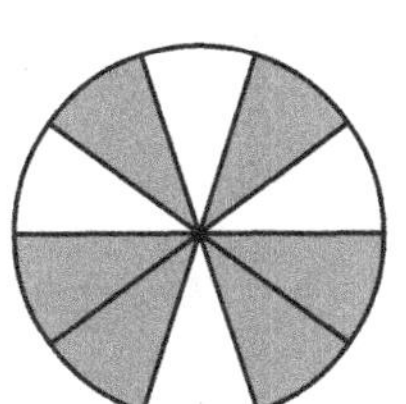

c 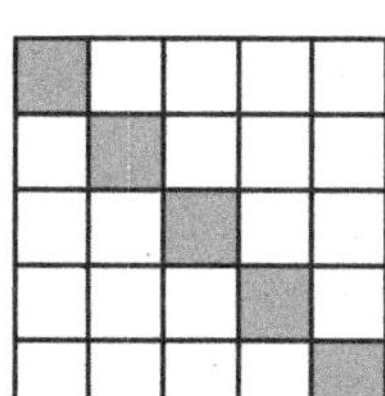

d 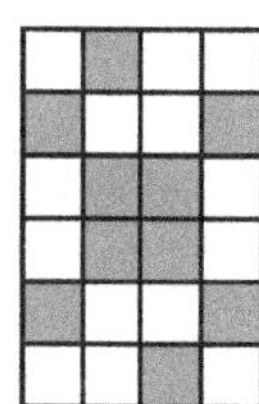

e 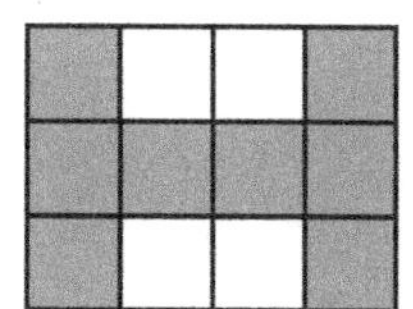

f 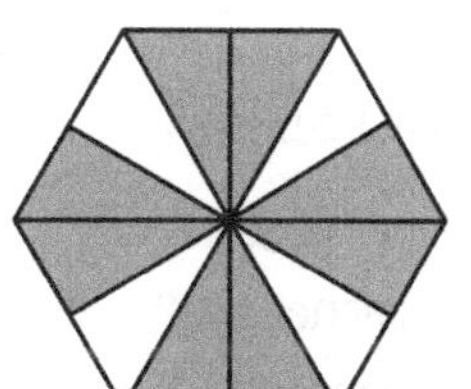

g 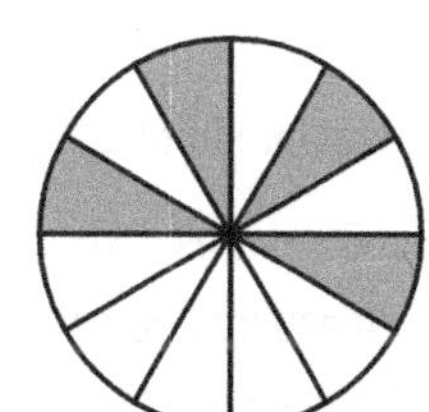

h 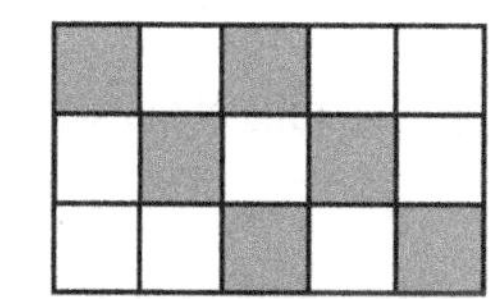

Compare ratios by converting units and solve ratio problems

Help Box

Before ratios can be compared, the quantities or amounts must be converted to the same unit.

Example: 1 cm to 1 m has to be converted to 1 cm to 100 cm which can then be expressed as 1:100; 8 hours to 1 day has to be converted to 8 hours to 24 hours which can then be expressed as 8:24 and simplified to 1:3.

1 Change the amounts in each ratio to the same unit and then write as a ratio in its simplest form.

a	10 cm to 1 m	**b**	1 m to 1 km	**c**	5 cm to 5 mm
d	20t to K1	**e**	K5 to 50t	**f**	K2 to 40t
g	$\frac{1}{2}$ kg to 900 g	**h**	1.25 kg to 500 g	**i**	750 g to 1.5 kg
j	2 hours to 1 day	**k**	1 day to 18 hours	**l**	3 hours to 1 day
m	2 L to 800 mL	**n**	250 mL to 5 L	**o**	100 mL to 1.5 L
p	30 minutes to 3 hours	**q**	2 hours to 20 minutes	**r**	15 minutes to 2.5 hours
s	400 kg to 1 tonne	**t**	650 kg to 1 tonne	**u**	1.2 tonne to 300 kg

2 Convert these map scales to the same unit of measurement and write the scale as a ratio in its simplest form.

a	1 cm = 1 km	**b**	1 cm = 5 m	**c**	2 cm = 5 m	**d**	5 cm = 1 km
e	2 cm = 10 m	**f**	2 cm = 1 km	**g**	5 cm = 2 km	**h**	3 cm = 30 m
i	10 cm = 1 km	**j**	10 cm = 100 m	**k**	5 cm = 0.5 km	**l**	2 cm = 0.25 km
m	3 cm = 1.5 km	**n**	0.5 cm = 0.5 m	**o**	20 cm = $\frac{1}{2}$ km	**p**	10 cm = 2 km

3 Rensie has K3 and Yerema has 70t.

a What is the ratio of Rensie's money to Yerema's money?

b What is the ratio of Yerema's money to Rensie's money?

4 A mixture is made of $\frac{3}{4}$ L of water and 200 mL of syrup.

a What is the ratio of water to syrup?

b What is the ratio of syrup to water?

5 At the trade store, the ratio of sales of packets of sweet biscuits to packets of savoury biscuits is 5:7.

a If 60 packets of sweet biscuits are sold, how many packets of savoury biscuits are sold?

b If 63 packets of savoury biscuits are sold, how many packets of sweet biscuits are sold?

6 The ratio of the sides for a rectangular set of dishes of different sizes is 3:2.

a If the shorter side of a dish measures 20 cm, what is the longer side?

b If the longer side of a dish measures 48 cm, what is the shorter side?

c If the shorter side of a dish measures 18 cm, what is the longer side?

d If the longer side of a dish measures 36 cm, what is the shorter side?

7.1.7 Use directed numbers in concrete problems

Compare and order directed numbers

Remember

Directed numbers are positive and negative whole numbers including zero. A minus sign is used to show negative numbers. A plus sign can be used to show positive numbers but is not necessary. A number line shows the distance of a number from zero in both positive and negative directions.

1 Use the number line to help you fill the gaps with < or >.

a	+6 □ −3	**b**	−7 □ −5	**c**	+2 □ +5	**d**	3 □ −1	**e**	−4 □ +4
f	8 □ +2	**g**	−6 □ −2	**h**	−11 □ 0	**i**	7 □ +9	**j**	−12 □ 4
k	0 □ 10	**l**	+6 □ −9	**m**	−2 □ −3	**n**	+4 □ 11	**o**	−5 □ 5
p	−9 □ 0	**q**	−10 □ −11	**r**	12 □ +9	**s**	−8 □ −9	**t**	5 □ +8

2 Imagine that the number line extends in both directions and write < or > to make each statement true.

a	+23 □ 25	**b**	−18 □ 20	**c**	−37 □ −20	**d**	34 □ 45	**e**	−19 □ +15
f	41 □ +39	**g**	+33 □ −30	**h**	−17 □ 28	**i**	46 □ +49	**j**	−34 □ −48
k	26 □ −16	**l**	+42 □ −14	**m**	−55 □ −49	**n**	43 □ −36	**o**	+49 □ 0
p	−37 □ −47	**q**	18 □ +8	**r**	+57 □ −57	**s**	−38 □ −40	**t**	+26 □ 19

3 Write the number from each of these groups that has the least value.

a	75, −73, +70	**b**	+55, −53, −58	**c**	80, −81, +78	**d**	−40, +65, −76
e	−96, 98, −93	**f**	+39, −39, +40	**g**	−14, 53, 47	**h**	−43, 0, +16
i	+27, −35, −28	**j**	96, +87, −89	**k**	+19, 16, −32	**l**	−14, +11, −17

4 Arrange the following sets of numbers in order from highest value to lowest value.

a	−8, +12, −1, −15, +3	**b**	+31, +17, −5, 0, −22	**c**	27, −18, −9, 23, +14
d	38, −2, +16, −13, −7	**e**	−3, −43, +1, 35, −21	**f**	+48, −59, −53, 42, 36
g	−29, 32, +29, −26, 25	**h**	−5, 0, +6, −6, −12	**i**	−40, 55, −51, +44, 43
j	95, +88, −67, −72, +91	**k**	+63, −60, −65, 72, 59	**l**	−36, −48, −41, 38, 47
m	−10, 0, −5, −14, +8	**n**	77, −73, +81, 87, −82	**o**	+28, −45, +56, −42, 49

+20 +19 +18 +17 +16 +15 +14 +13 +12 +11 +10 +9 +8 +7 +6 +5 +4 +3 +2 +1 0 −1 −2 −3 −4 −5 −6 −7 −8 −9 −10 −11 −12 −13 −14 −15 −16 −17 −18 −19 −20

Use directed numbers to solve problems

Use the vertical number line on the left of this page to help you solve the following word problems.

1 Joe entered the lift on the 4th floor. What floor did he finish on if he went down the following number of floors?

a 9 floors **b** 5 floors **c** 2 floors **d** 7 floors

2 What floor would a lift end at if it begins at the following floors and goes up or down as described?

a the 4th floor below ground level and goes up 8 floors
b the 2nd floor above ground level and goes down 5 floors
c the 3rd floor below ground level and goes down 3 floors
d ground level and goes down 7 floors
e the 2nd floor below ground level and goes up 4 floors
f the 5th floor above ground level and goes down 9 floors
g the 7th floor above ground level and goes down 10 floors
h the 4th floor above ground level and goes down 7 floors

3 The temperature was –3° in the morning but had risen 10° by lunchtime. What was the temperature at lunchtime?

4 What would the temperature be if it starts at the following temperatures and rises or falls as described?

a –2°C and rises 9°C **b** 15°C and falls 12°C **c** 5°C and falls 8°C
d –6°C and rises 5°C **e** –10°C and rises 12°C **f** 4°C and falls 11°C

5 Mona had a credit card debt of K186. Over a twelve-month period, the amounts shown in the chart below were added and deducted. Copy and calculate the amount owing at the end of each transaction.

Month	Transactions	Amount owing
		K186
January	Paid off K55	
February	Paid off K27	
March	Added extra debt of K45	
April	Paid off K25	
May	Paid off K36	
June	Paid off K30	
July	Added extra debt of K44	
August	Paid off K32	
September	Added extra debt of K53	
October	Paid off K18	
November	Paid off K36	
December	Paid off K24	

7.1.8 Use positive indices greater than the power of 1

Use positive indices

Remember

Index numbers (indices) are used to show that a number is multiplied by itself a certain number of times.
For example, in $2^5 = 2 \times 2 \times 2 \times 2 \times 2$, the 5 is the index number that shows 2 is multiplied by itself 5 times.

1 Use index notation to record these multiplications.

a	$7 \times 7 \times 7 \times 7 \times 7 \times 7$	**b**	$4 \times 4 \times 4 \times 4 \times 4$
c	$2 \times 2 \times 2 \times 2 \times 2 \times 2 \times 2 \times 2$	**d**	$5 \times 5 \times 5 \times 5 \times 5 \times 5 \times 5$
e	$8 \times 8 \times 8 \times 8 \times 8$	**f**	$3 \times 3 \times 3 \times 3 \times 3 \times 3 \times 3 \times 3 \times 3$
g	$6 \times 6 \times 6 \times 6$	**h**	$9 \times 9 \times 9$
i	$4 \times 4 \times 4 \times 4 \times 4 \times 4 \times 4$	**j**	$2 \times 2 \times 2 \times 2 \times 2$
k	$3 \times 3 \times 3 \times 3 \times 3 \times 3 \times 3 \times 3$	**l**	$6 \times 6 \times 6 \times 6 \times 6 \times 6 \times 6 \times 6 \times 6 \times 6$
m	$10 \times 10 \times 10 \times 10$	**n**	$8 \times 8 \times 8 \times 8 \times 8 \times 8 \times 8$
o	$5 \times 5 \times 5 \times 5 \times 5 \times 5$	**p**	$11 \times 11 \times 11 \times 11 \times 11$

2 Write each of these as multiplications.

a	5^3	**b**	7^4	**c**	2^2	**d**	9^7	**e**	3^9	**f**	4^6
g	2^5	**h**	8^3	**i**	6^4	**j**	5^7	**k**	9^2	**l**	10^4
m	1^9	**n**	3^5	**o**	7^8	**p**	11^4	**q**	6^9	**r**	4^7

3 Calculate answers to these additions and subtractions.

a	$3^3 + 2^5$	**b**	$10^3 - 4^4$	**c**	$5^2 + 5^3$	**d**	$8^5 - 10^2$
e	$6^4 + 9^3$	**f**	$5^5 - 7^3$	**g**	$11^3 + 3^5$	**h**	$12^2 - 4^3$
i	$2^9 + 1^8$	**j**	$6^3 - 3^4$	**k**	$7^4 + 7^2$	**l**	$6^5 - 9^4$
m	$3^5 + 5^4$	**n**	$10^4 - 10^3$	**o**	$9^4 + 5^3$	**p**	$8^3 - 7^3$

4 Calculate answers to these multiplications.

a	$5^3 \times 2^3$	**b**	$3^3 \times 4^2$	**c**	$10^3 \times 2^2$	**d**	$6^3 \times 5^2$
e	$4^4 \times 10^2$	**f**	$4^5 \times 2^4$	**g**	$9^3 \times 4^2$	**h**	$7^2 \times 8^5$
i	$11^3 \times 3^3$	**j**	$5^2 \times 9^4$	**k**	$6^5 \times 2^4$	**l**	$2^9 \times 10^3$
m	$12^3 \times 8^2$	**n**	$7^5 \times 6^2$	**o**	$3^8 \times 4^3$	**p**	$9^1 \times 5^5$

5 Fill the gaps with =, < or >.

a $35 \square 6^2$ b $10^3 \square 500$ c $1500 \square 3^7$ d $8^4 \square 2000$

e $5^3 \square 10^2$ f $12^3 \square 4^4$ g $7^3 \square 5^6$ h $11^3 \square 8^4$

i $2^9 \square 3^8$ j $6^5 \square 5^6$ k $9^3 \square 4^7$ l $3^9 \square 7^5$

m $4^5 \square 5^4$ n $7^3 \square 3^7$ o $10^4 \square 9^5$ p $4^6 \square 2^8$

6 Calculate and write the largest number in each group.

a 3^3 5^2 4^2 b 5^3 10^2 2^6 c 4^4 3^5 7^3 d 2^8 3^5 4^3

e 10^3 6^4 4^5 f 8^4 5^5 7^5 g 2^9 6^3 3^4 h 5^4 7^3 9^3

i 8^3 4^4 3^5 j 11^4 12^3 10^4 k 9^5 7^6 5^7 l 10^6 8^8 4^{10}

m 6^2 7^3 3^9 n 4^5 6^3 7^2 o 4^6 10^2 9^3 p 3^8 5^4 2^9

7

5^4	6^7	4^2	10^4	3^5	9^3	12^3	6^5	3^8	2^7

a Which is the largest number?

b Which is the smallest number?

c Which numbers equal less than 1000?

d Which numbers equal more than 5000?

e Which number equals 128?

f Which numbers are between 600 and 2000?

g Which two numbers total 857?

h Find the difference between the two largest numbers.

8

6^6	4^4	9^4	3^6	3^7	11^3	8^5	2^9	5^3	10^3

a Which is the smallest number?

b Which is the largest number?

c Which numbers equal more than 1000?

d Which number equals 1000?

e Which numbers are between 500 and 1500?

f Which numbers are less than 500?

g Which two numbers total 3518?

h Find the difference between the largest number and the smallest number.

9 Find answers to these.

a $5^2 + 6^2 + 2^7$ b $7^3 + 7^2 + 7^4$ c $2^4 + 3^5 + 4^3$ d $2^8 + 1^3 + 7^4$

e $3^3 + 10^3 + 5^3$ f $4^4 + 6^3 + 3^4$ g $8^2 + 9^3 + 11^2$ h $5^4 + 3^7 + 2^6$

i $10^2 + 5^5 + 7^2$ j $3^4 + 2^3 + 8^3$ k $8^5 + 12^3 + 6^2$ l $4^8 + 9^2 + 2^5$

10 Fill the gaps with < or >.

a $4^3 + 4^2 \square 3^5 + 2^4$ b $7^3 - 2^5 \square 10^3 - 8^3$ c $3^3 \times 11^2 \square 6^4 + 5^5$

d $9^4 - 2^7 \square 7^2 \times 4^3$ e $3^3 + 5^4 \square 6^5 - 8^4$ f $7^5 - 4^6 \square 9^2 \times 5^4$

g $3^4 + 5^2 \square 4^4 - 5^2$ h $2^6 \times 2^2 \square 10^2 + 6^3$ i $3^7 + 2^6 \square 2^3 \times 3^5$

j $7^4 - 4^2 \square 8^2 + 8^3$ k $9^4 - 4^3 \square 6^2 \times 7^2$ l $4^5 + 8^4 \square 3^9 - 4^7$

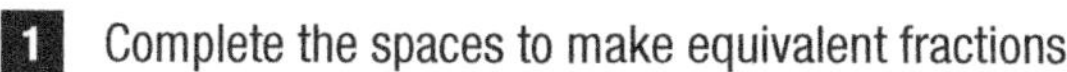

Assessment Number and Application

1 Complete the spaces to make equivalent fractions.

a $\frac{8}{12} = \frac{2}{\square}$ b $\frac{3}{7} = \frac{\square}{28}$ c $\frac{8}{36} = \frac{2}{\square}$ d $\frac{7}{12} = \frac{\square}{48}$

2 Write these fractions in their simplest form.

a $\frac{12}{16}$ b $\frac{8}{18}$ c $\frac{20}{25}$ d $\frac{9}{63}$ e $\frac{6}{20}$ f $\frac{10}{22}$

3 Convert these mixed numbers to improper fractions.

a $4\frac{1}{5}$ b $2\frac{5}{8}$ c $5\frac{3}{4}$ d $1\frac{2}{3}$ e $3\frac{5}{6}$ f $6\frac{2}{3}$

4 Convert these improper fractions to mixed numbers.

a $\frac{16}{3}$ b $\frac{29}{6}$ c $\frac{17}{7}$ d $\frac{31}{9}$ e $\frac{22}{5}$ f $\frac{37}{8}$

5 Solve these fraction additions and subtractions. Write the answers as mixed numbers where possible, and reduce all fractions to their simplest form.

a $\frac{7}{12} + \frac{7}{12}$ b $\frac{8}{9} - \frac{5}{9}$ c $2\frac{2}{3} + 1\frac{5}{6}$ d $4\frac{3}{10} - 2\frac{4}{5}$

e $2\frac{1}{8} + 3\frac{2}{5}$ f $3\frac{3}{7} - 1\frac{2}{3}$

6 Solve these multiplications and divisions and write the answers in their simplest form.

a $\frac{2}{7} \times \frac{2}{3}$ b $\frac{7}{8} \div \frac{1}{3}$ c $9 \times 2\frac{3}{4}$ d $1\frac{7}{10} \div 3$

e $2\frac{4}{5} \times 1\frac{3}{7}$ f $5\frac{1}{4} \div 1\frac{5}{6}$

7 Solve these fraction word problems.

a A bike was travelling at $22\frac{3}{4}$ km per hour. At this speed, how many kilometres would be travelled in $4\frac{5}{6}$ hours?

b Isaac walked $7\frac{5}{8}$ km to the market. Peter walked $4\frac{2}{3}$ km to the same market. How much further did Isaac walk than Peter?

c One suitcase weighs $19\frac{2}{5}$ kg and another suitcase weighs $21\frac{3}{8}$ kg. What is the total weight of the two suitcases?

8 Use $<$ or $>$ between each pair of decimals to make true statements.

a 4.51 $\square$ 4.6 b 8.07 $\square$ 8.077 c 3.183 $\square$ 3.18 d 2.72 $\square$ 2.025

e 5.9 $\square$ 5.89 f 6.41 $\square$ 6.411

Assessment Number and Application

9 Order each set of numbers from largest to smallest.

a 4.2 4.275 4.02 4.21 4.07 4.033 4.16 4.025

b 7.86 7.9 7.08 7.816 7.082 7.8 7.19 7.09

c 2.45 2.05 2.56 2.4 2.005 2.425 2.004 2.5

10 Copy and complete this chart.

Decimal number	Rounded to one decimal place	Rounded to two decimal places	Rounded to nearest whole number
4.116			
8.552			
5.073			
2.855			
6.007			

11 Solve these decimal additions and subtractions.

a 23.76 + 18.95 b 5.853 + 78.37 c 156.8 + 68.65 d 39.629 + 146.48

e 81.06 – 35.88 f 153.4 – 75.59 g 203.03 – 89.6 h 126.14 – 68.7

12 Solve these decimal multiplications.

a 36.4 × 100 b 57.23 × 10 c 8.757 × 30 d 81.45 × 60

e 75.3 × 7 f 5.865 × 24 g 64.48 × 5.3 h 159.7 × 8.6

13 Solve these decimal divisions.

a 76.12 ÷ 10 b 1538.6 ÷ 100 c 816.8 ÷ 20 d 137.35 ÷ 50

e 809.4 ÷ 6 f 866.52 ÷ 18 g 161.82 ÷ 5.8 h 216.24 ÷ 3.4

14 In a group of 10 students, 5 wore red shirts, 2 wore blue shirts and 3 wore white shirts. Write each of the following as a ratio and a fraction.

a the number of red shirts to the number of white shirts

b the number of blue shirts to the number of red shirts

c the number of white shirts to the total number of shirts

Assessment

Number and Application

15 Solve these decimal word problems.

- **a** The total weight of 15 boxes is 131.55 kg. What is the average weight of each box?
- **b** What would be the cost of 3.7 kg of mince at K8.55 per kilogram?
- **c** Jay worked a part-time job for three weeks. The first week he earned K227.75, the second week K186.40 and the third week K197.85. How much did he earn over the three weeks?

16 Write each of these ratios in their simplest form.

a 4:10 **b** 18:14 **c** 5:20 **d** 60:10 **e** 8:36 **f** 50:15

17 Change the amounts in each ratio to the same unit and then write as a ratio in its simplest form.

- **a** 20 cm to 1 m
- **b** 6 hours to 1 day
- **c** K10 to 50t
- **d** 4 L to 400 mL
- **e** 300 g to 1 kg
- **f** 8 months to 2 years

18 Use $<$ or $>$ to fill the gaps.

- **a** -2 □ $+7$
- **b** 15 □ -5
- **c** -8 □ -11
- **d** $+4$ □ -9
- **e** -32 □ -30
- **f** -21 □ 0
- **g** 44 □ $+42$
- **h** -39 □ 36

19 Order these sets of numbers from lowest to highest value.

- **a** –6, +10, –14, –8, +17
- **b** 5, –11, 0, –16, +9
- **c** +17, –17, 20, +6, –12
- **d** +63, –50, –62, 57, –23
- **e** 38, –35, –41, +37, –29
- **f** –86, 80, +71, -92, –87

20 What would the bank balance be if it starts at the following amounts and increases or decreases as described?

- **a** K35 and decreases K42
- **b** –K22 and increases K15
- **c** –K14 and decreases K17
- **d** –K8 and increases K63

21 Use index notation to record these multiplications.

- **a** $9 \times 9 \times 9 \times 9 \times 9 \times 9 \times 9$
- **b** $12 \times 12 \times 12 \times 12 \times 12$
- **c** $25 \times 25 \times 25$
- **d** $16 \times 16 \times 16 \times 16 \times 16 \times 16$

22 Solve these calculations.

- **a** $2^4 + 3^3$
- **b** $5^3 - 6^2$
- **c** $4^4 \times 2^3$
- **d** $9^3 + 3^5$
- **e** $11^3 - 2^5$
- **f** $6^3 \times 2^7$
- **g** $3^7 + 8^2$
- **h** $10^4 - 6^5$

23 Fill the gaps with $<$ or $>$.

- **a** $4^2 + 5^2$ □ $10^2 - 2^6$
- **b** $3^4 \times 5^2$ □ $9^3 + 8^3$
- **c** $5^5 - 7^2$ □ $2^4 \times 2^3$

Assessment Number and Application

Multiple choice test for 7.1.3 and 7.1.4

1 Which decimal is equivalent to $\frac{27}{100}$?

a 0.027 b 0.27 c 2.7 d 2.07

2 Which decimal is equivalent to $\frac{65}{1000}$?

a 0.65 b 0.605 c 0.065 d 6.5

3 Which decimal is equivalent to $\frac{16}{25}$?

a 0.16 b 0.64 c 0.6 d 0.1625

4 Which decimal is equivalent to $\frac{11}{20}$?

a 0.55 b 0.11 c 0.055 d 5.5

5 Which decimal is equivalent to $\frac{6}{7}$?

a 0.67 b 6.7 c 0.8 d 0.86

6 Which fraction is equivalent to 0.057?

a $\frac{57}{100}$ b $\frac{57}{1000}$ c $\frac{57}{10}$ d $5\frac{7}{10}$

7 Which fraction is equivalent to 3.71?

a $3\frac{71}{1000}$ b $3\frac{71}{10}$ c $3\frac{7}{10}$ d $3\frac{71}{100}$

8 Which percentage is equivalent to $\frac{87}{100}$?

a 8.7% b 87% c 870% d 0.87%

9 Which percentage is equivalent to 0.09?

a 9% b 90% c 0.9% d 0.09%

10 Which percentage is equivalent to $\frac{3}{5}$?

a 35% b 6% c 60% d 3.5%

11 Which decimal is equivalent to 47%?

a 4.7 b 0.47 c 4.07 d 0.047

12 Which decimal is equivalent to $17\frac{1}{2}\%$?

a 17.5 b 1.75 c 0.175 d 17.05

13 Which fraction is equivalent to 130%?

a $\frac{3}{10}$ b $\frac{3}{100}$ c $1\frac{3}{100}$ d $1\frac{3}{10}$

14 Which fraction is equivalent to $22\frac{1}{2}\%$?

a $\frac{2}{5}$ b $\frac{45}{100}$ c $\frac{45}{2}$ d $\frac{9}{40}$

15 What is 5% of K120?

a K6 b K12 c K1.20 d K12.50

16 What is $12\frac{1}{2}\%$ of K350?

a K4375 b K43.75 c K400 d K437.50

17 What is K160 increased by 20%?

a K180 b K165 c K192 d K128

18 What is 860 g increased by 15%?

a 875 g b 900 g c 950 g d 989 g

19 What is K375 increased by 30%?

a K487.50 b K405 c K112.50 d K450

Strand **Space and Shape**

7.2.2 Use appropriate metric units in calculations

Convert mixed units to assist calculations

Remember

When comparing or doing calculations with lengths, it is important that the lengths are converted to the same unit.

1 Write the lengths that are longer than $\frac{1}{2}$ metre.

35 cm 6000 mm 1000 mm 250 cm 68 cm
450 mm 70 mm 48 cm 55 cm 510 mm
46 cm 550 mm 495 mm 60 cm 506 mm

2 Write the lengths that are shorter than 2 metres.

150 cm 1600 mm 2500 mm 15 cm 250 cm
210 cm 20 cm 5000 mm 500 mm 2250 mm
197 cm 1950 mm 203 cm 2005 mm 1995 mm

3 Convert each set of measurements to the same unit and then write the longest measurement in each set.

- **a** 3.5 cm, 38 mm, 0.3 cm
- **b** 15 cm, 1.5 cm, 15 mm
- **c** 1.2 m, 125 cm, 1000 mm
- **d** 0.8 cm, 9 mm, 9.5 mm
- **e** 56 mm, 0.58 m, 5.8 cm
- **f** 4.25 m, 400 cm, 4200 mm
- **g** 8.9 cm, 895 mm, 0.89 m
- **h** 410 mm, 4.1 cm, 0.4 m
- **i** 2.1 m, 215 cm, 2105 mm
- **j** 10.5 mm, 1.5 cm, 0.15 cm
- **k** 0.7 km, 640 m, 7100 mm
- **l** 67.5 m, 0.6 km, 50 000 cm

4 Write each set of measurements in order from longest to shortest.

- **a** 17 m, 32 mm, 0.4 km, 2.5 cm
- **b** 50 mm, 0.52 km, 50 cm, 50 m
- **c** 0.65 m, 65 km, 6.5 m, 6500 cm
- **d** 87 cm, 76 m, 0.76 km, 8700 cm
- **e** 350 mm, 0.035 m, 3.55 cm, 0.35 km
- **f** 0.93 km, 9300 m, 9300 cm, 9.3 m

5 Solve these problems by converting the lengths in each problem to the same unit.

- **a** A marathon runner had run 8700 metres of a 25 km race. How far did she still have to run?
- **b** A builder has three pieces of timber measuring 1600 mm, 0.213 m and 98 cm. What is the total of the three lengths?
- **c** Sam runs 17 laps of an oval. If each lap is a distance of 275 m, how many kilometres does Sam run?
- **d** A piece of rope is 9.57 m long. If it is cut into eleven equal parts, how many centimetres long is each piece of rope?
- **e** If a swimmer swims 43 laps of a 50 m pool, how many kilometres will he have swum?

7.2.3 Investigate and measure the circumference of circles

Calculate circumference

The circumference of the circle below can be calculated using the formula $C = \pi D$ where C is circumference, D is diameter and π represents 3.14.

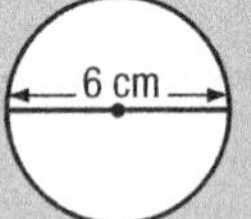

The diameter is 6 cm so $C = 3.14 \times 6$ cm. The circumference is 18.84 cm.

The circumference of the circle below can be calculated using the formula $C = 2\pi r$ where C is circumference, r is radius and π represents 3.14.

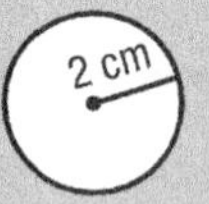

The radius is 2 cm so $C = 2 \times 3.14 \times 2$ cm. The circumference is 12.56 cm.

Find the circumference of each of the following circles by measuring either the radius or diameter, and using that measurement to calculate circumference.

1

2

3

4

5

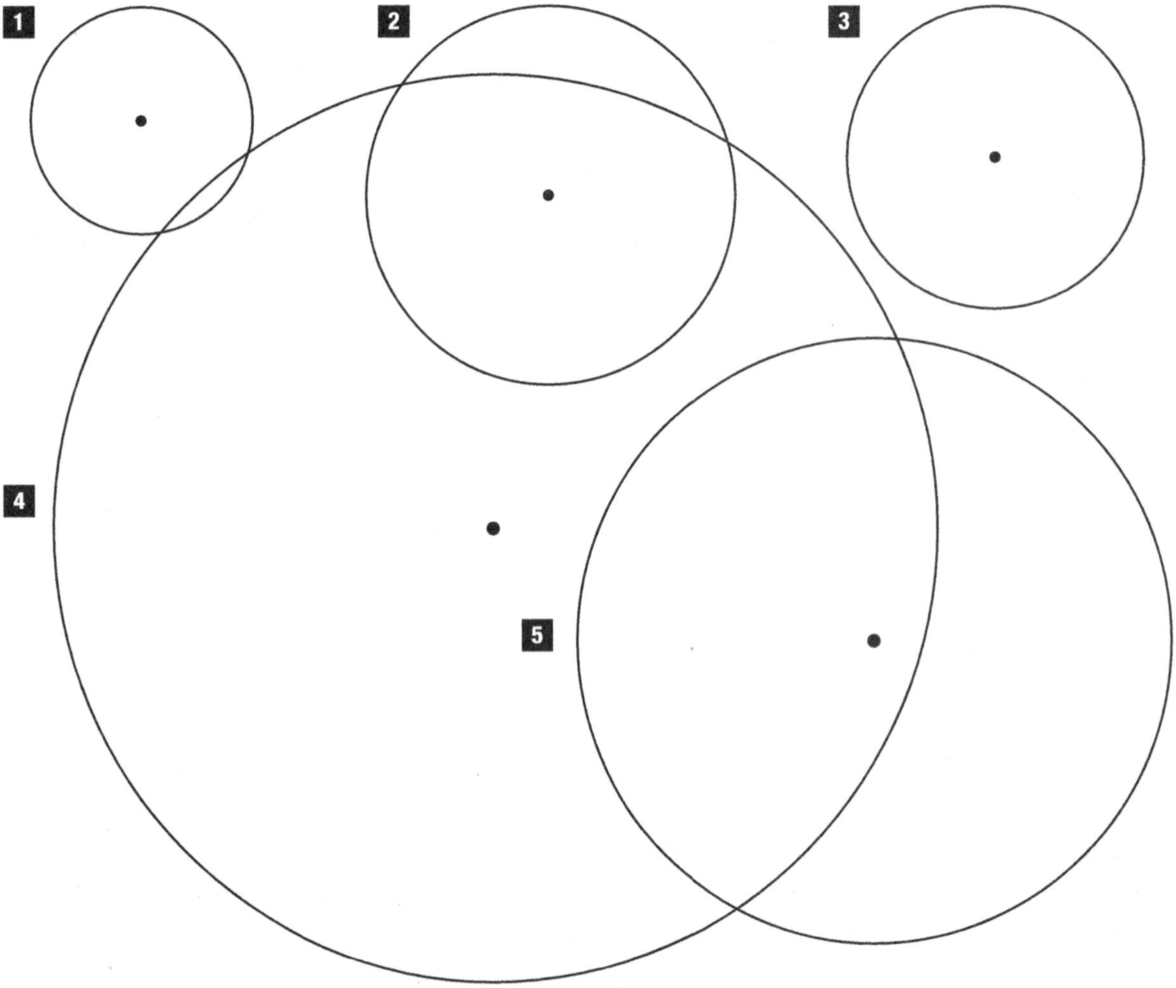

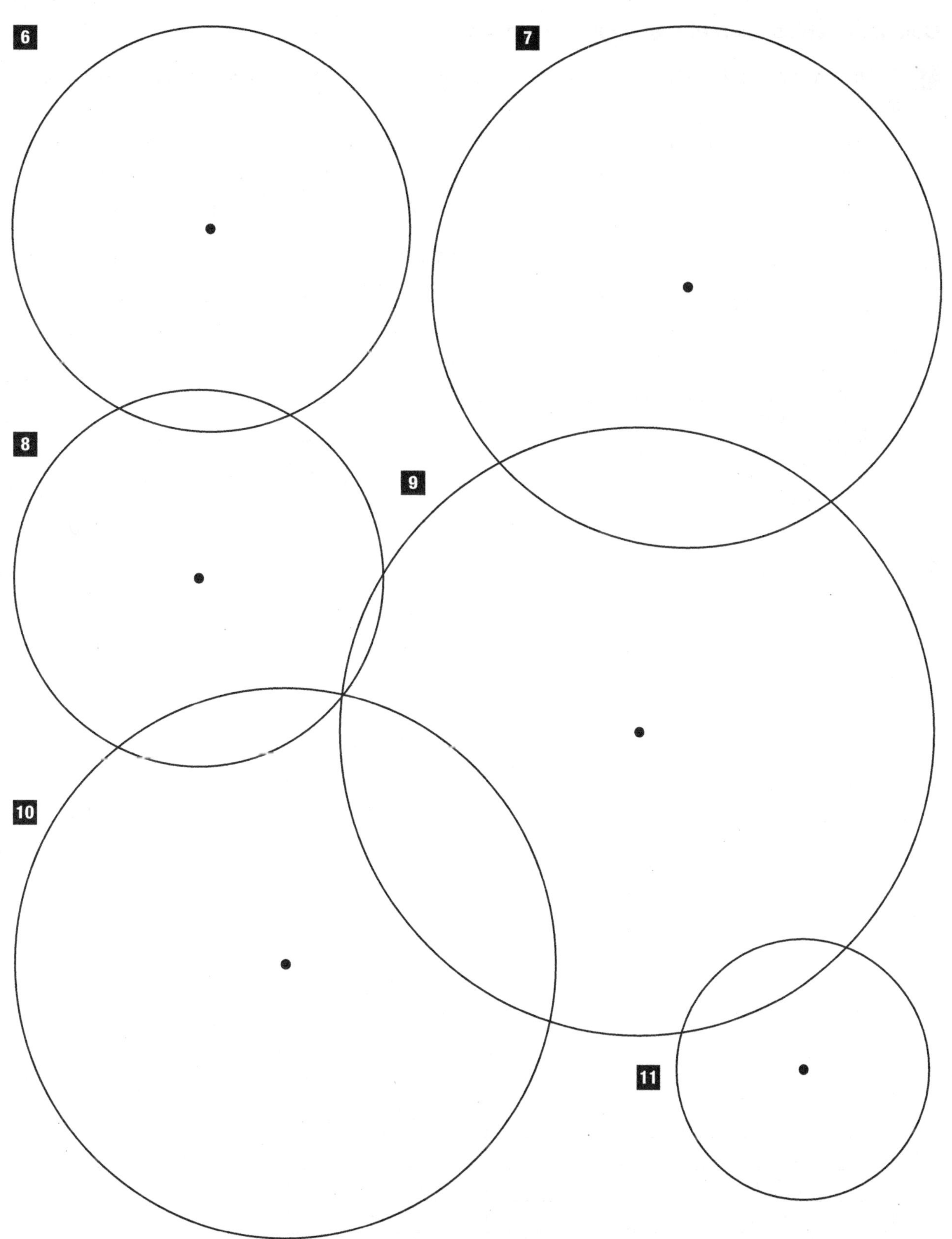
6
7
8
9
10
11

Calculate radius, diameter and circumference

1 Find the circumference of circles that have the following diameters. Where necessary, round answers to two decimal places.

a	20 cm	**b**	15 cm	**c**	10 cm	**d**	16 cm	**e**	23 cm	**f**	11 cm
g	18 cm	**h**	25 cm	**i**	30 cm	**j**	50 cm	**k**	17.5 cm	**l**	9.25 cm
m	7.75 cm	**n**	12.5 cm	**o**	13.2 cm	**p**	21.7 cm	**q**	19.3 cm	**r**	8.9 cm

2 Find the circumference of circles that have the following radii. Where necessary, round answers to two decimal places.

a	5 cm	**b**	13 cm	**c**	4 cm	**d**	17 cm	**e**	8 cm	**f**	14 cm
g	16 cm	**h**	9.5 cm	**i**	12.5 cm	**j**	7.5 cm	**k**	5.75 cm	**l**	11.6 cm
m	8.25 cm	**n**	14.9 cm	**o**	6.8 cm	**p**	16.75 cm	**q**	10.35 cm	**r**	3.85 cm

Remember

If you know the circumference of a circle, you can use the formula C ÷ 3.14 to calculate the diameter. You can then find the radius by dividing the diameter by 2.

3 Copy and complete this chart.

	Radius	Diameter	Circumference
a			43.96 cm
b			72.22
c	6.35 cm		
d		15.4 cm	
e		8.2 cm	
f			19.782 cm
g	11.7 cm		
h		27.3 cm	
i	9.18 cm		
j			30.144 cm
k		33.5 cm	
l			83.21 cm
m	4.85 cm		
n		12.7 cm	
o	5.25 cm		
p			64.684 cm
q			72.534 cm
r	14.65 cm		
s	8.45 cm		

7.2.5 Investigate area rules for quadrilaterals

Calculate area of rectangles

Remember

To calculate the area of a rectangle, multiply the length by the width (L × W).

1 Use the given measurements to calculate the area of each rectangle.

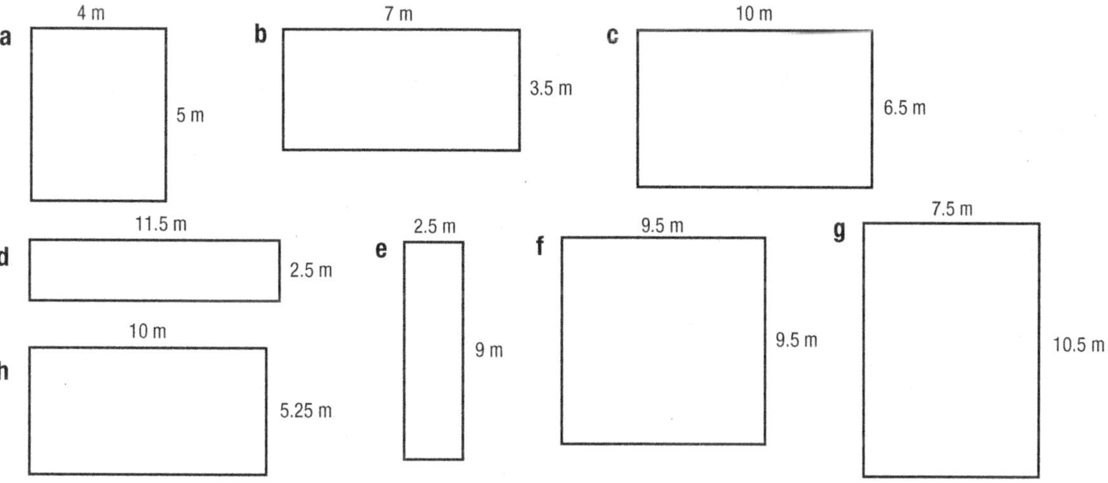

2 Measure the length and width of each rectangle and use the measurements to calculate the area.

a

b

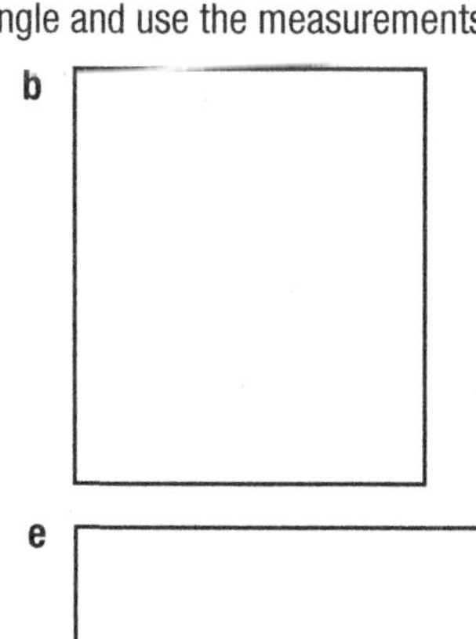

c

d

f

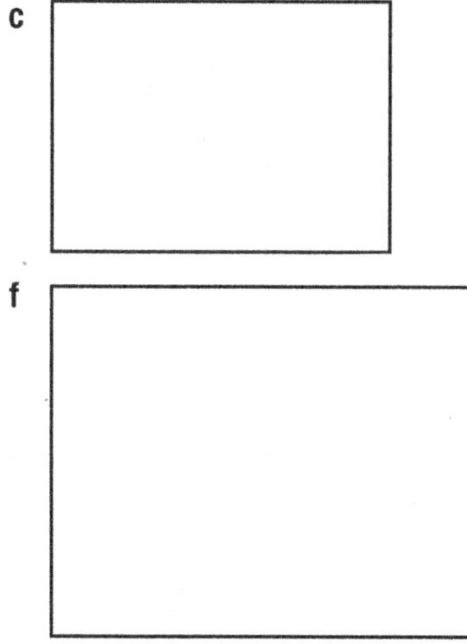

e

g

Calculate area of triangles

Remember

To calculate the area of a triangle, halve the product of its height and base. Another way to say this is: multiply the base by the height and then divide the answer by 2: ($\frac{b \times h}{2}$).

You will also need to use this formula when calculating the area of kites and composite shapes that contain triangles or kites.

Use the given measurements to calculate the area of each triangle.

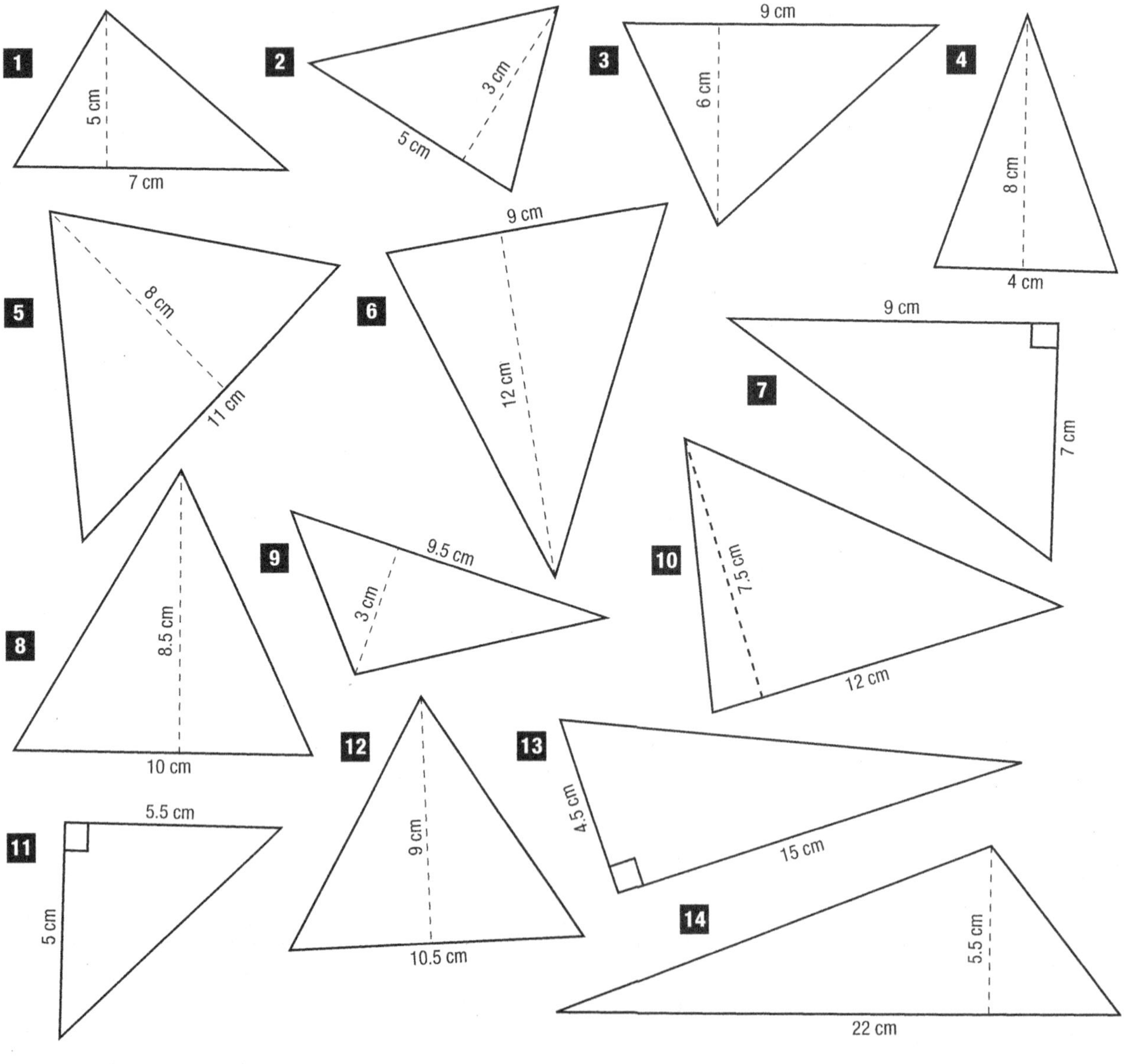

1 5 cm; 7 cm

2 3 cm; 5 cm

3 9 cm; 6 cm

4 8 cm; 4 cm

5 8 cm; 11 cm

6 9 cm; 12 cm

7 9 cm; 7 cm

8 8.5 cm; 10 cm

9 9.5 cm; 3 cm

10 7.5 cm; 12 cm

11 5.5 cm; 5 cm

12 9 cm; 10.5 cm

13 4.5 cm; 15 cm

14 5.5 cm; 22 cm

Calculate area of kites

Help Box

A kite is a shape made up of two triangles. To find its area, you need to find the area of each triangle and add the two areas together.

Example: The area of the smaller triangle in the kite is

$\frac{3 \text{ cm} \times 5 \text{ cm}}{2} = 15 \text{ cm} \div 2 = 7.5 \text{ cm}^2$

The area of the larger triangle in the kite is

$\frac{7 \text{ cm} \times 5 \text{ cm}}{2} = 35 \text{ cm} \div 2 = 17.5 \text{ cm}^2$

So, $7.5 \text{ cm}^2 + 17.5 \text{ cm}^2 = 25 \text{ cm}^2$

The area of the kite is 25 cm².

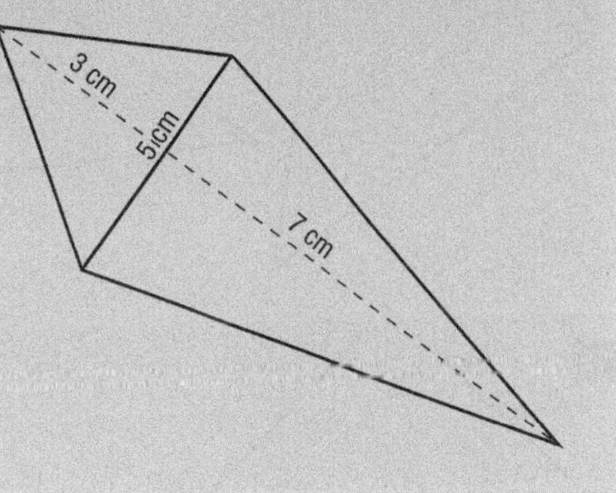

Use the given measurements to calculate the area of each kite.

1 6 cm, 4 cm, 2 cm

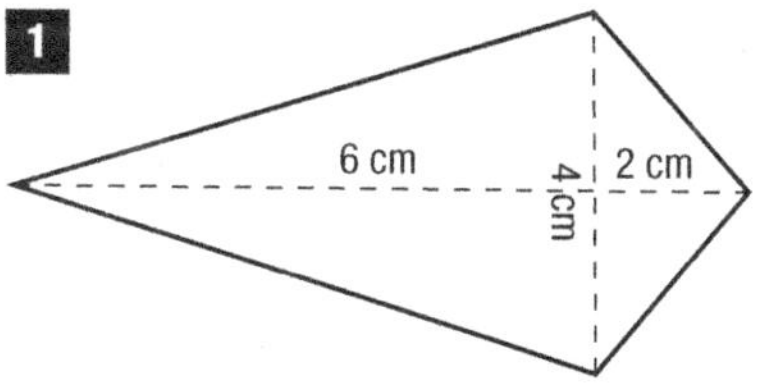

2 9 cm, 4 cm, 3 cm

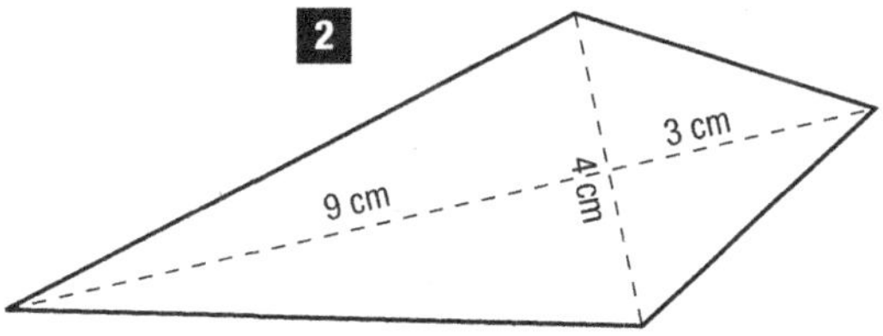

3 2 cm, 6 cm, 8 cm

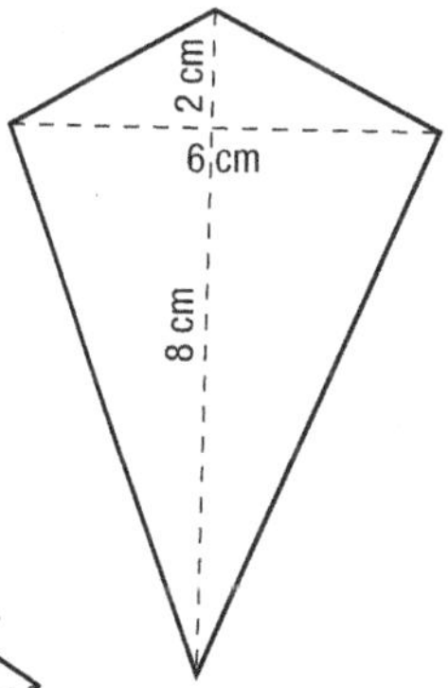

4 6 cm, 8 cm, 20 cm

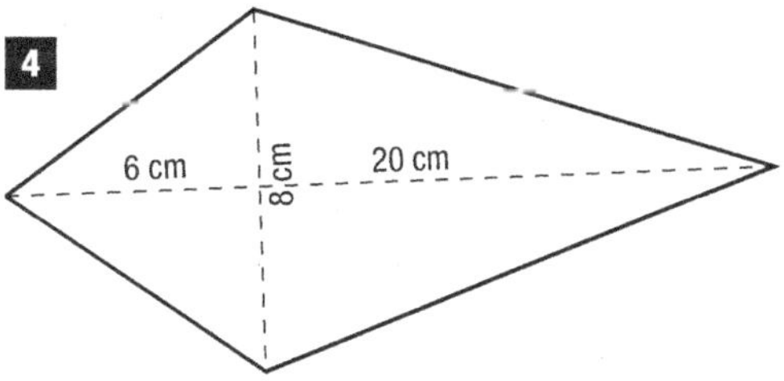

5 15 cm, 9 cm, 4 cm

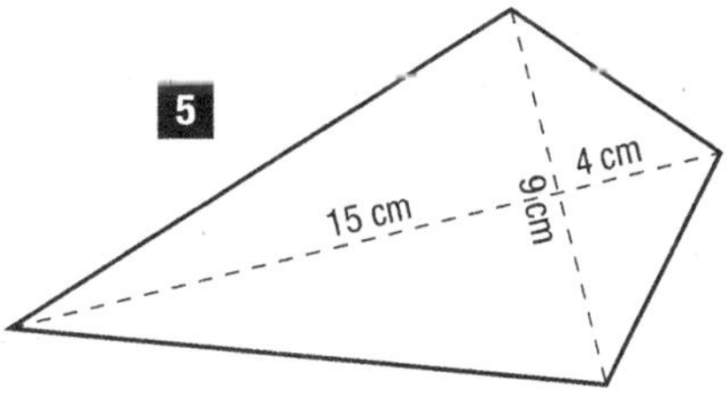

6 7 cm, 20 cm, 18 cm

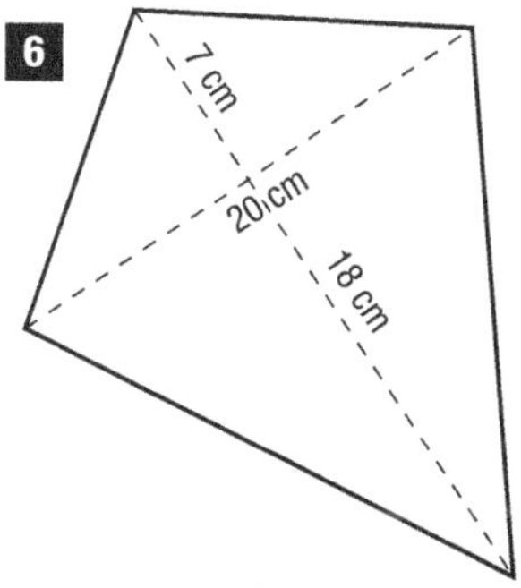

7 3 cm, 4.5 cm, 8 cm

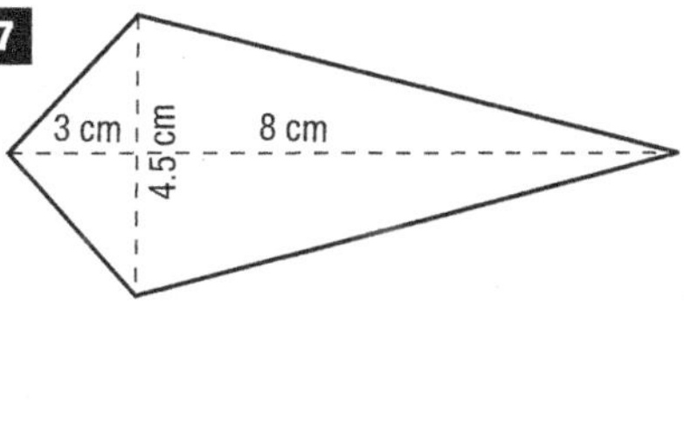

8 30 cm, 20 cm, 11.5 cm

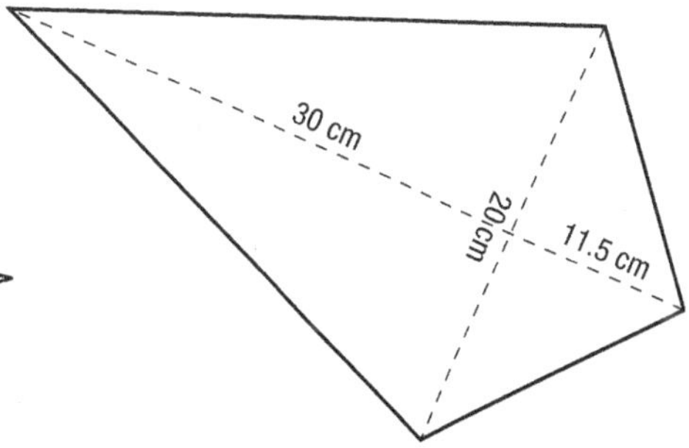

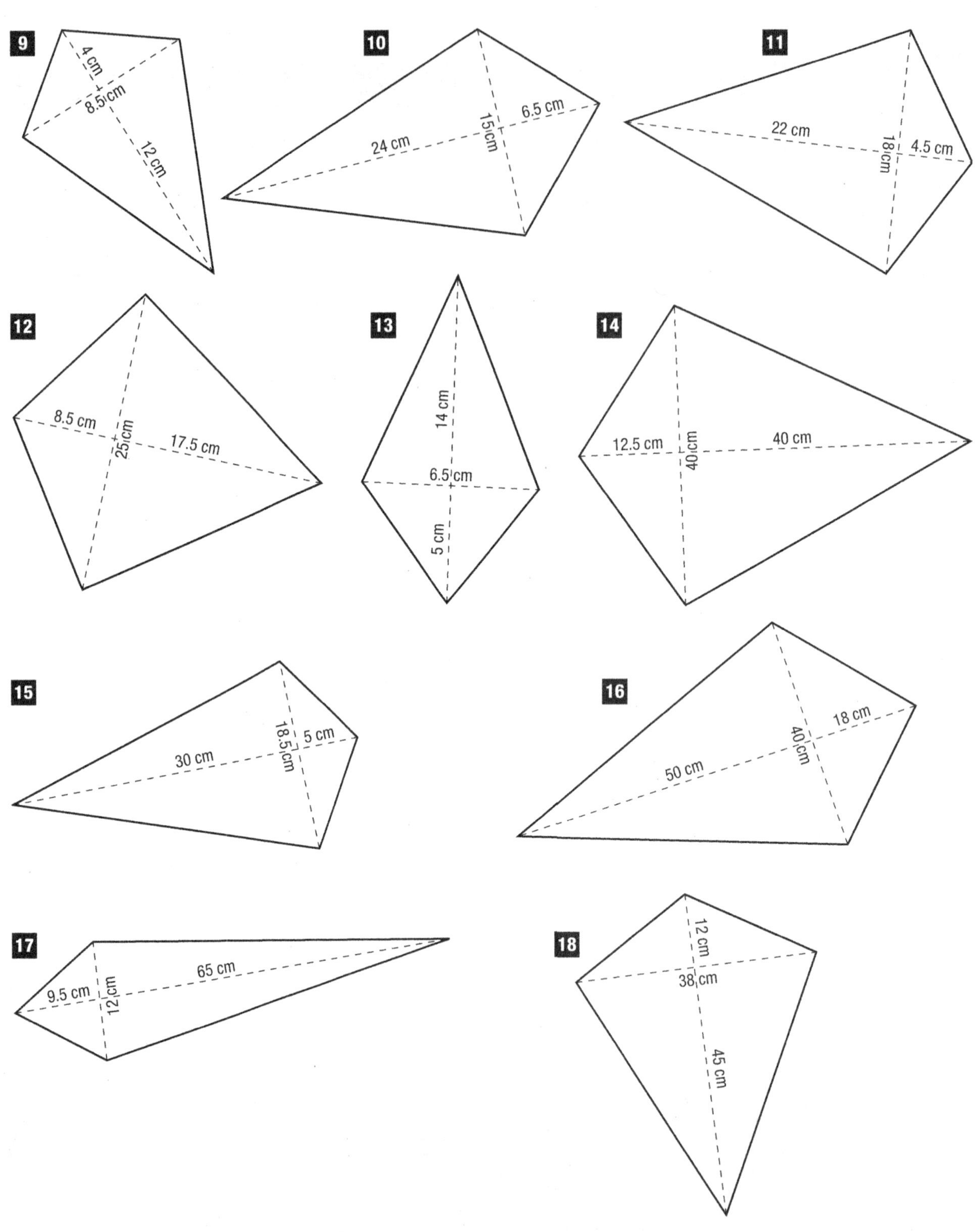
9
4 cm
8.5 cm
12 cm
10
6.5 cm
15 cm
24 cm
11
22 cm
18 cm
4.5 cm
12
8.5 cm
25 cm
17.5 cm
13
14 cm
6.5 cm
5 cm
14
12.5 cm
40 cm
40 cm
15
18.5 cm
5 cm
30 cm
16
18 cm
40 cm
50 cm
17
65 cm
9.5 cm
12 cm
18
12 cm
38 cm
45 cm

Calculate area of parallelograms

Help Box

The area of a parallelogram can be calculated by the same rule that is used to find the area of a rectangle: base × height, where height is the vertical height as shown in the diagram.

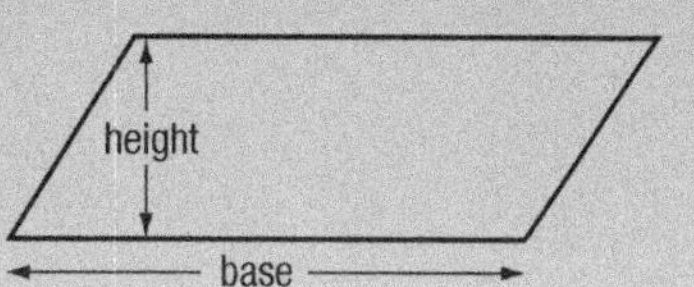

Use the given measurements to calculate the area of each parallelogram.

1 6 cm; 8 cm

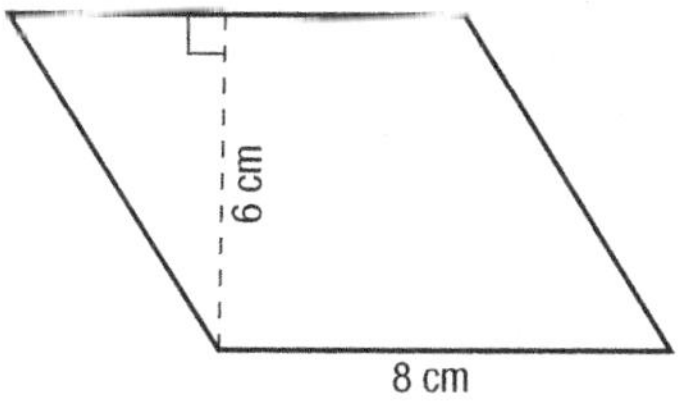

2 8 cm; 5 cm

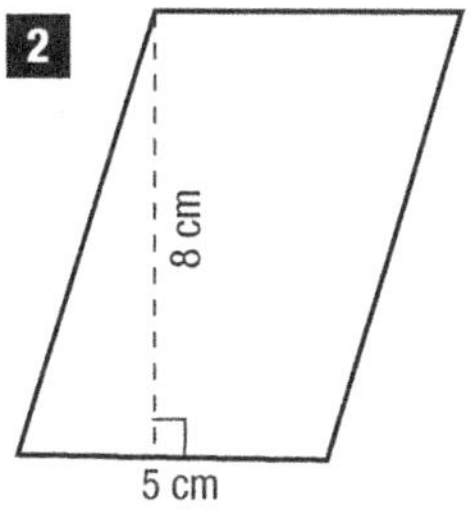

3 8 cm; 12 cm

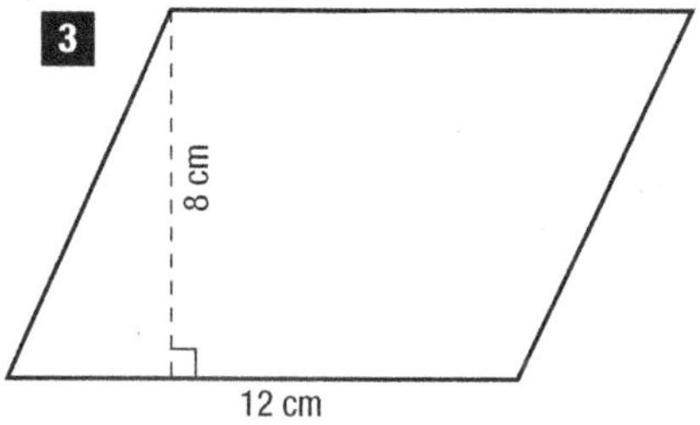

4 6.5 cm; 10 cm

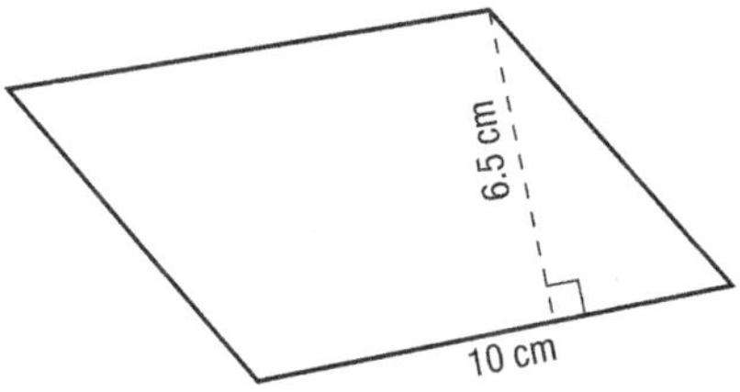

5 12 cm; 3.5 cm

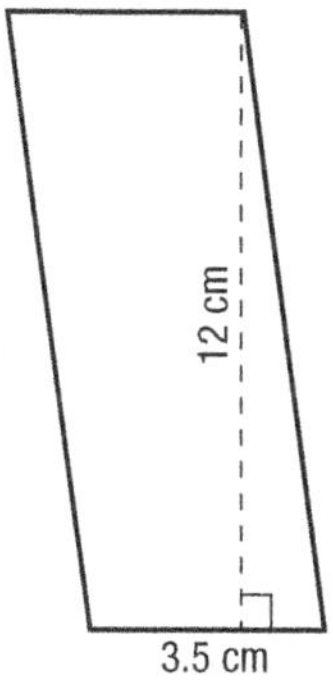

6 4.5 cm; 20 cm

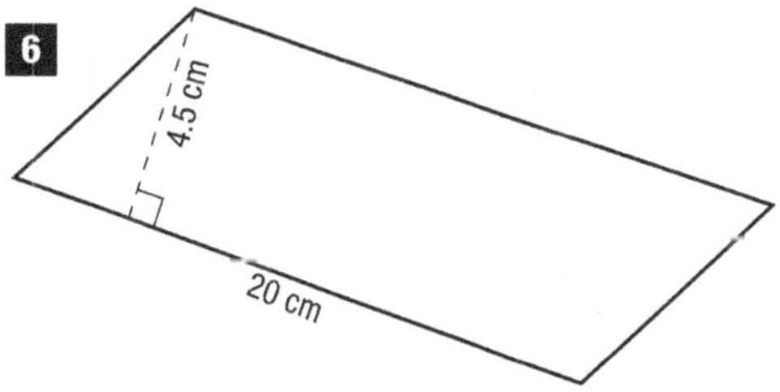

7 25 cm; 22.5 cm

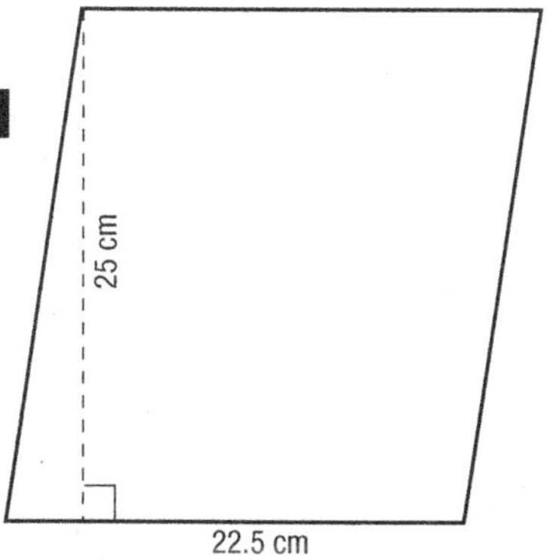

8 9.5 cm; 15 cm

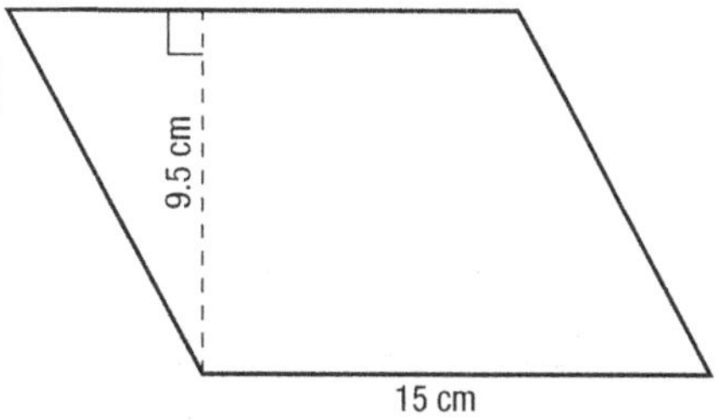

9 16 cm; 12 cm

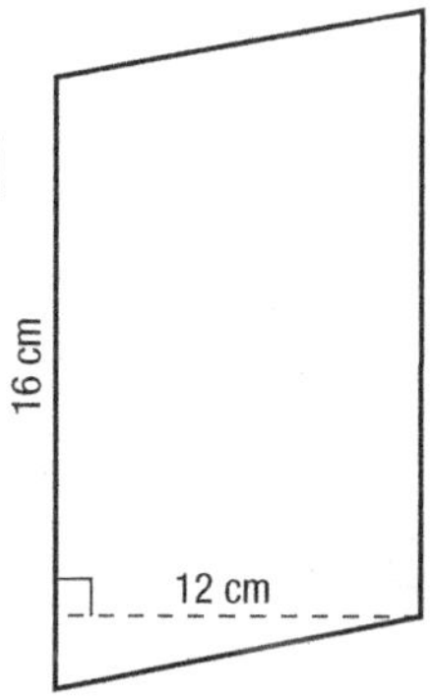

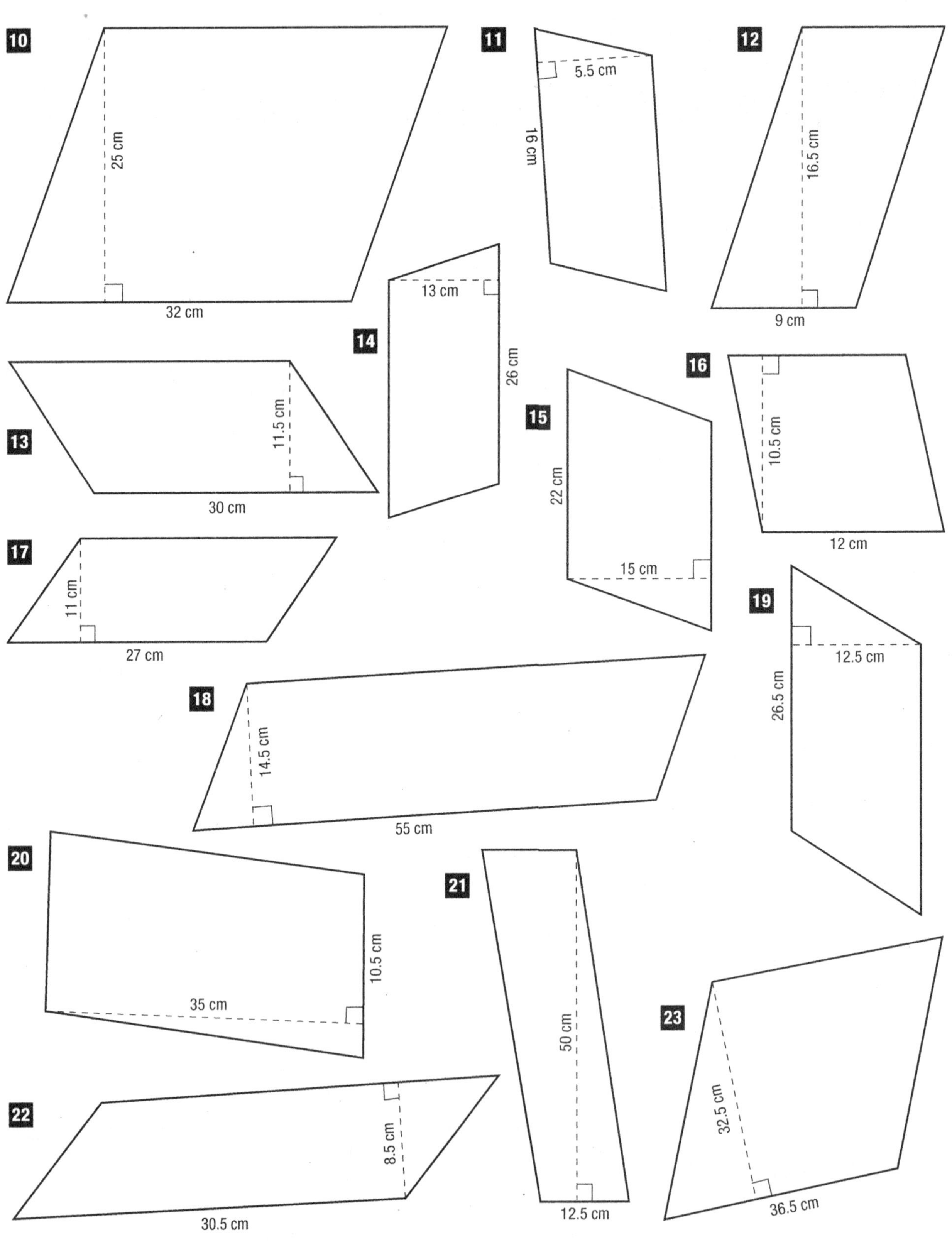
10
25 cm
32 cm
11
5.5 cm
16 cm
12
16.5 cm
9 cm
13
11.5 cm
30 cm
14
13 cm
26 cm
15
22 cm
15 cm
16
10.5 cm
12 cm
17
11 cm
27 cm
18
14.5 cm
55 cm
19
12.5 cm
26.5 cm
20
35 cm
10.5 cm
21
50 cm
12.5 cm
22
8.5 cm
30.5 cm
23
32.5 cm
36.5 cm

Calculate area of trapeziums

Help Box

The area of a trapezium can be calculated by first doubling the size of the trapezium to make a parallelogram as shown in the diagram. Then, find the area of the parallelogram and because it is twice the size of the trapezium, divide by 2.

This can be stated as $\frac{(a+b)h}{2}$

where a represents one of the parallel sides, b represents the other parallel side and h represents the vertical height.

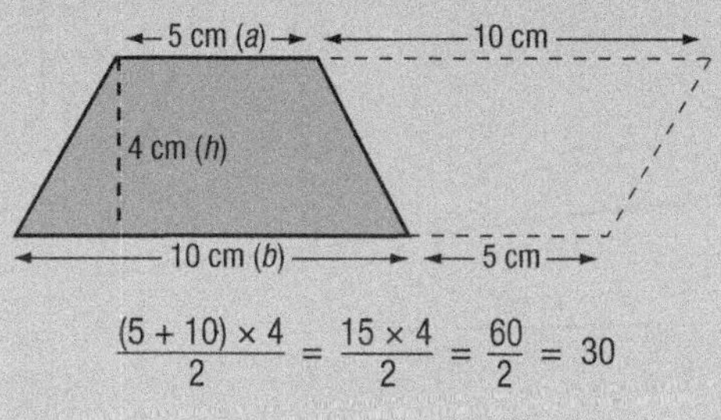

$$\frac{(5+10)\times 4}{2} = \frac{15\times 4}{2} = \frac{60}{2} = 30$$

The area is 30 cm^2.

Use the given measurements to calculate the area of each trapezium.

1 3 cm; 3 cm; 5 cm

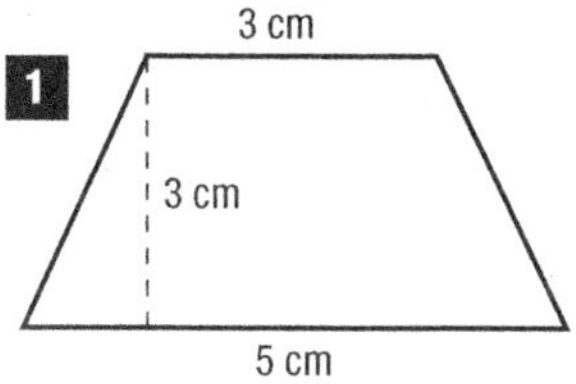

2 2 cm; 2 cm; 8 cm

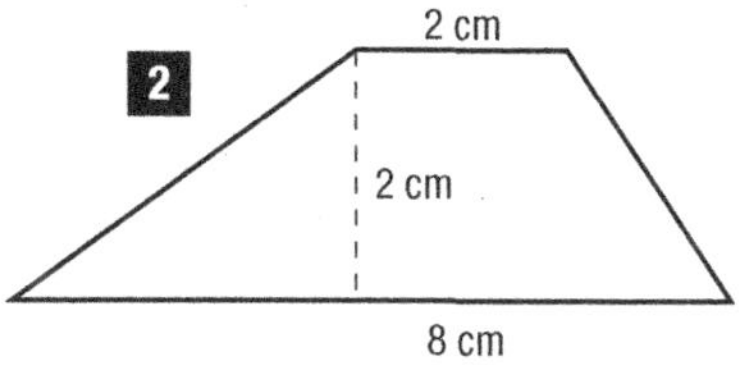

3 9 cm; 6 cm; 4 cm

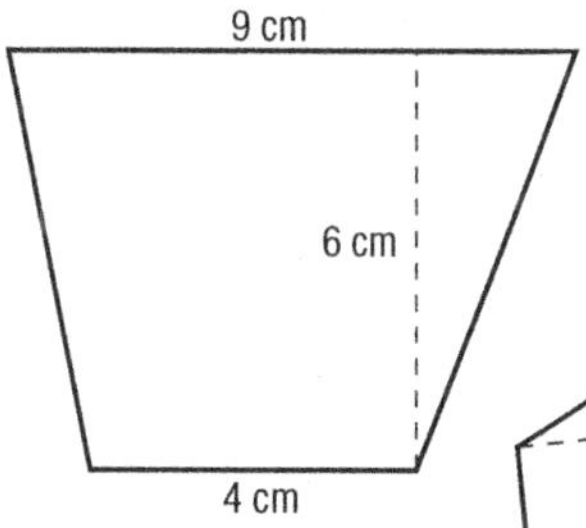

4 12 cm; 9 cm; 20 cm

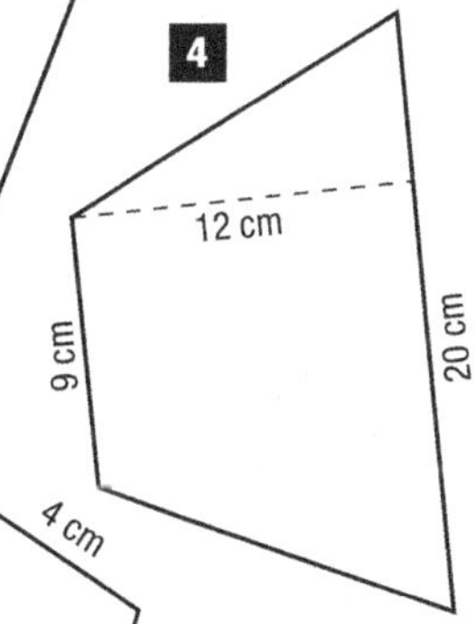

5 12 cm; 6 cm; 6 cm

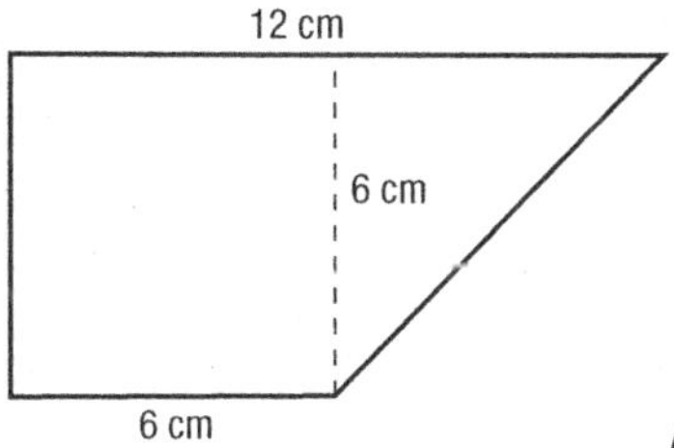

6 5 cm; 8 cm; 10 cm

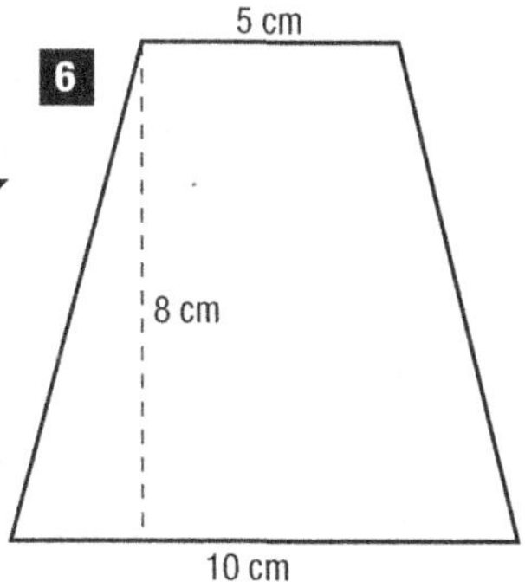

7 4 cm; 5 cm; 9 cm

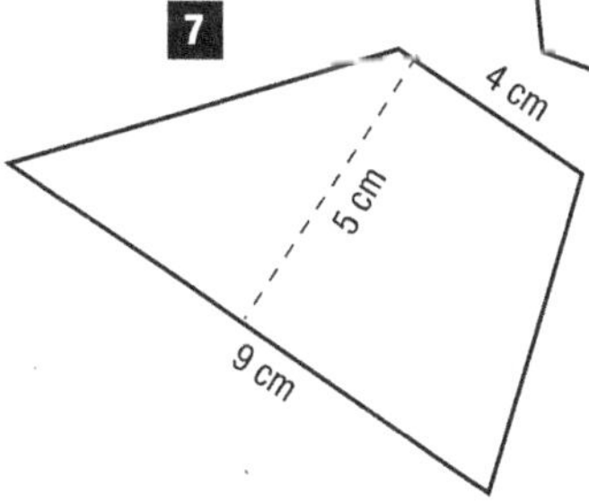

8 11 cm; 20 cm; 16 cm

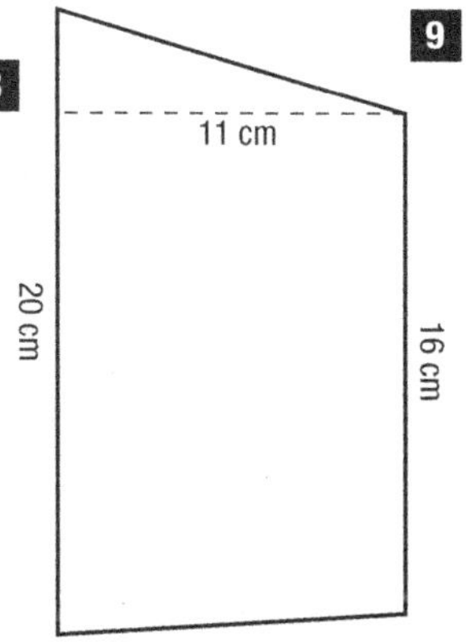

9 18 cm; 15 cm; 5 cm

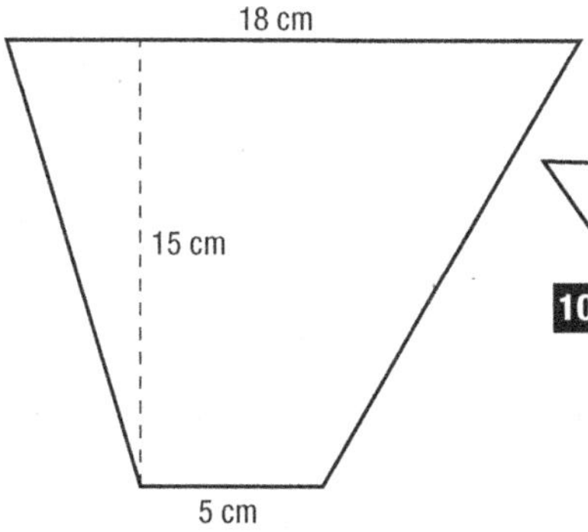

10 25 cm; 15 cm; 15 cm

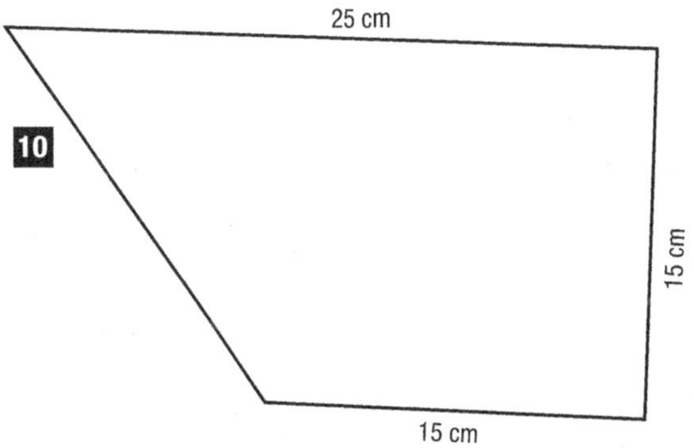

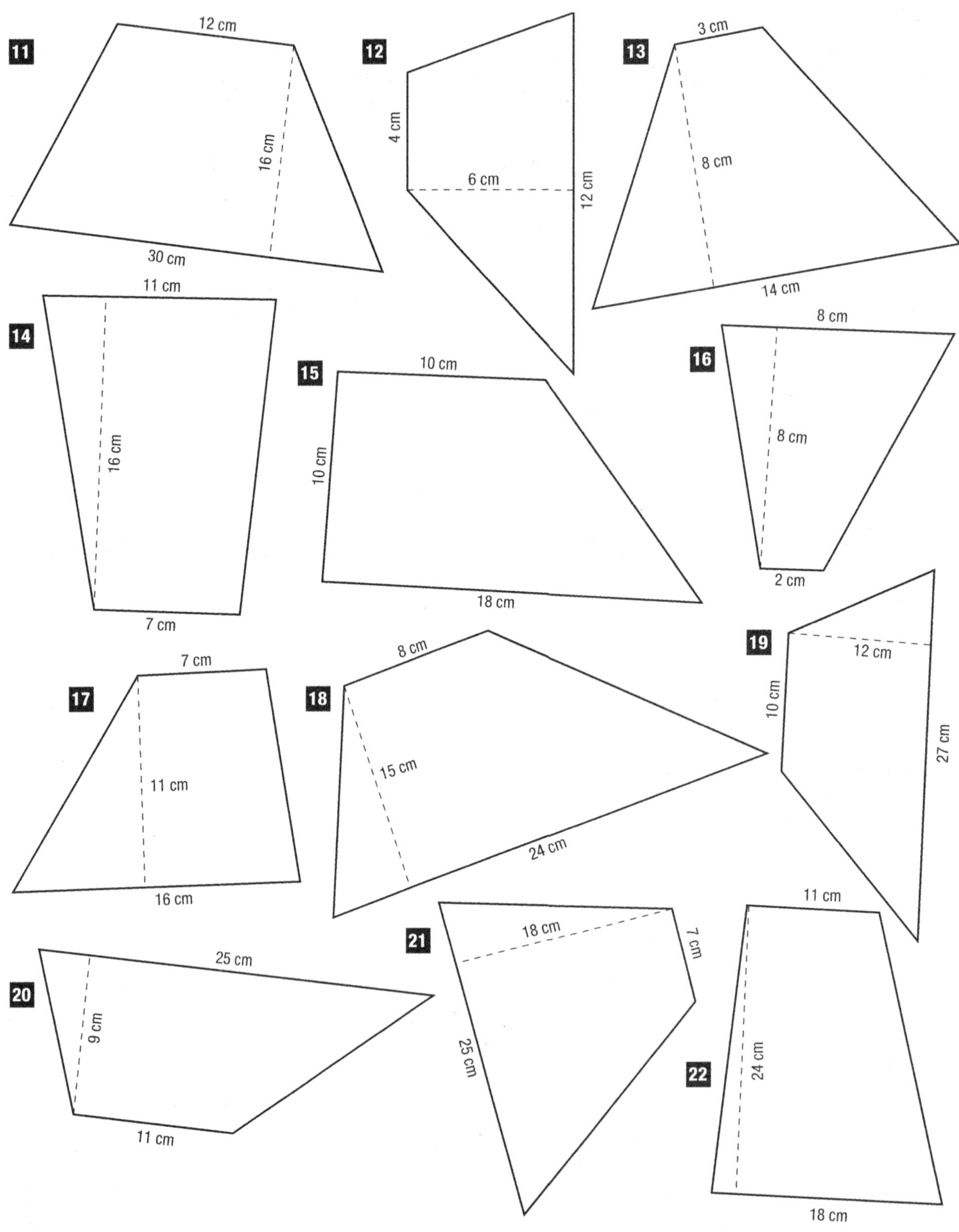
11
12 cm
16 cm
30 cm
12
4 cm
6 cm
12 cm
13
3 cm
8 cm
14 cm
14
11 cm
16 cm
7 cm
15
10 cm
10 cm
18 cm
16
8 cm
8 cm
2 cm
17
7 cm
11 cm
16 cm
18
8 cm
15 cm
24 cm
19
12 cm
10 cm
27 cm
20
25 cm
9 cm
11 cm
21
18 cm
7 cm
25 cm
22
11 cm
24 cm
18 cm

Calculate area of composite shapes

Remember

To calculate the area of a composite shape, split the shape into simpler shapes. Find the area of each simpler shape and then add them up to find the total area of the composite shape.

Use the given measurements to calculate the total shaded area of each shape.

1 10 m, 7 m, 12 m, 4 m, 5 m, 6 m

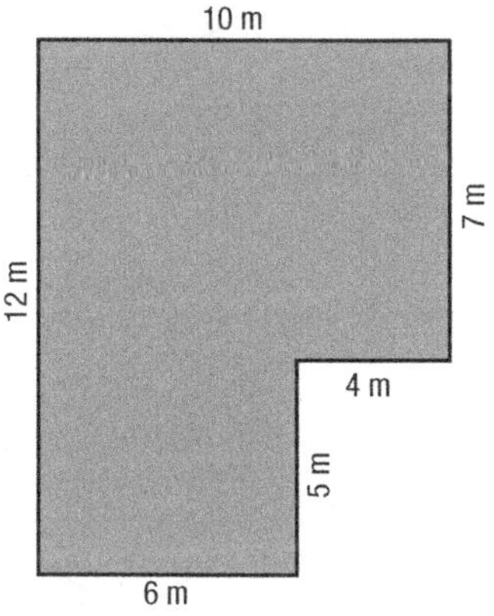

2 4 m, 4 m, 4 m, 5 m, 1 m, 9 m, 10 m, 5 m, 4 m, 5 m

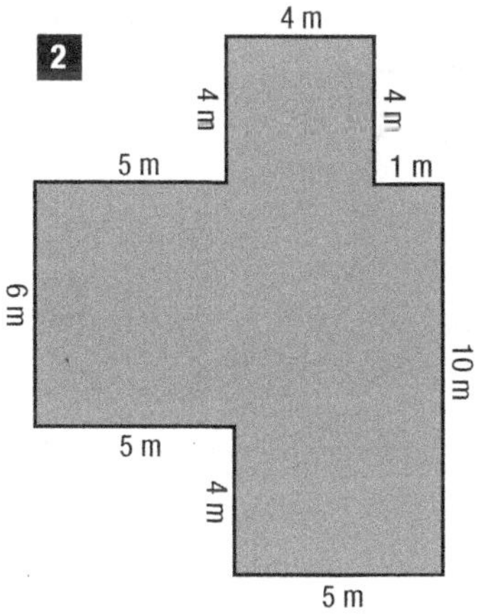

3 6 m, 3 m, 3 m, 6 m, 3 m, 9 m, 9 m, 15 m

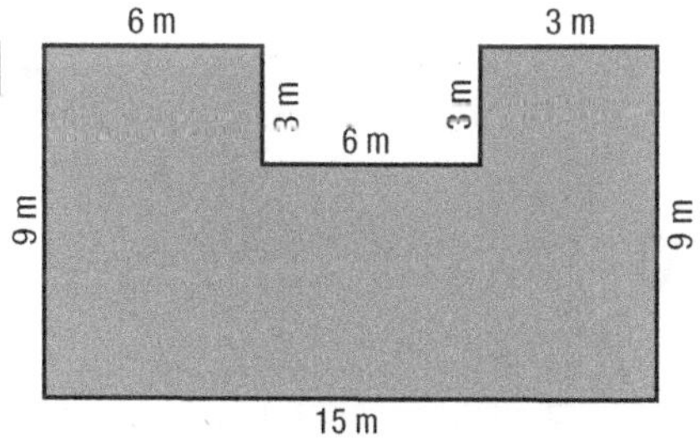

4 12 m, 5 m, 8 m, 5 m

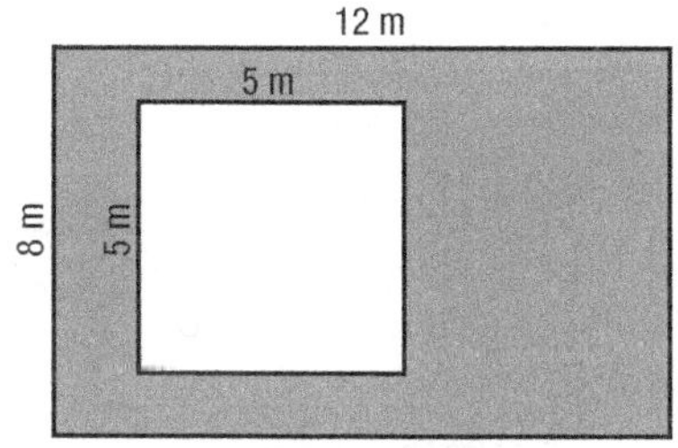

5 4 m, 4 m, 6 m, 12 m, 5 m, 8 m, 2 m, 10 m

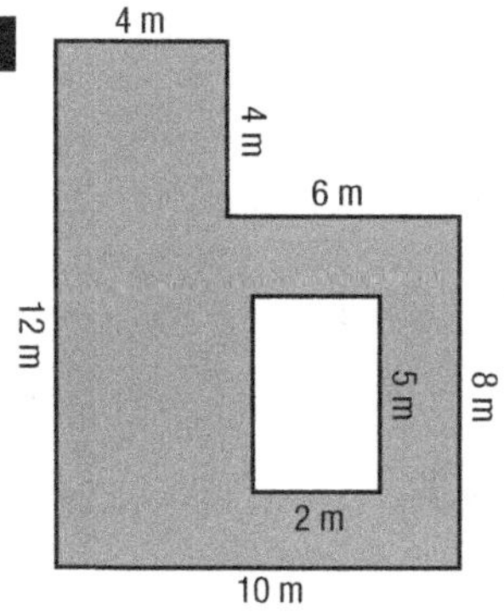

6 6 m, 9 m, 9 m, 4 m, 8 m, 6 m, 6 m

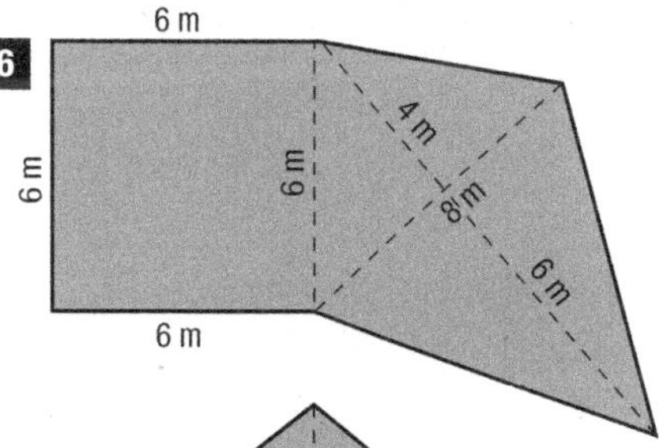

7 7 m, 3.5 m, 7 m, 7 m, 3.5 m, 7 m

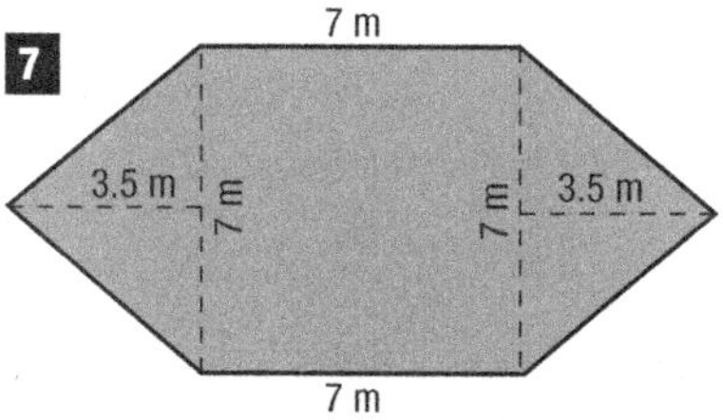

8 3 m, 8 m, 8 m, 10 m

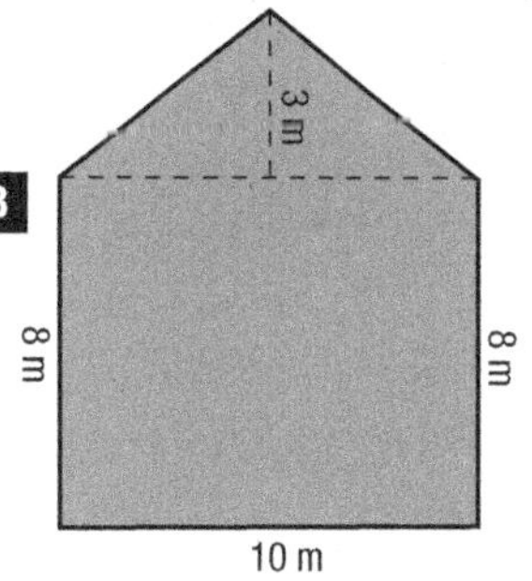

9 15 m, 7.5 m, 7.5 m, 15 m, 10 m

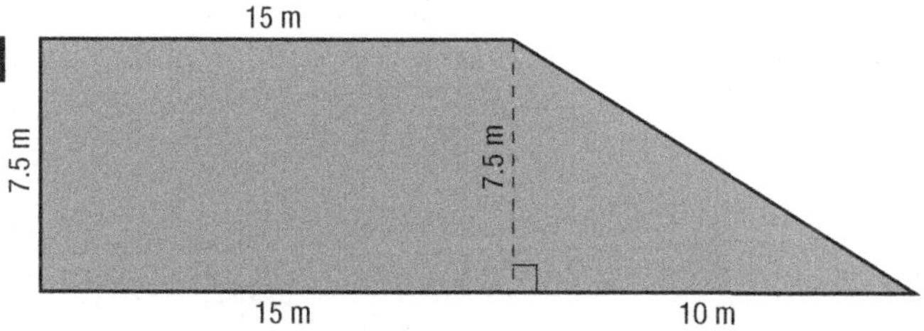

10 5 m, 8 m, 3 m, 4 m, 4 m, 3 m, 3 m, 10 m, 6 m, 12 m, 4 m, 3 m, 4 m, 8 m, 5 m

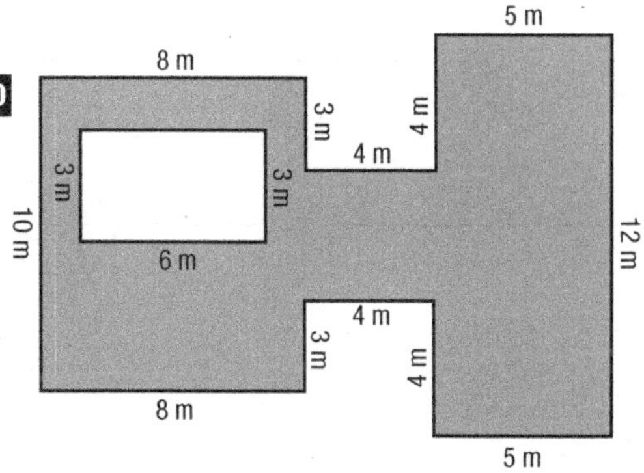

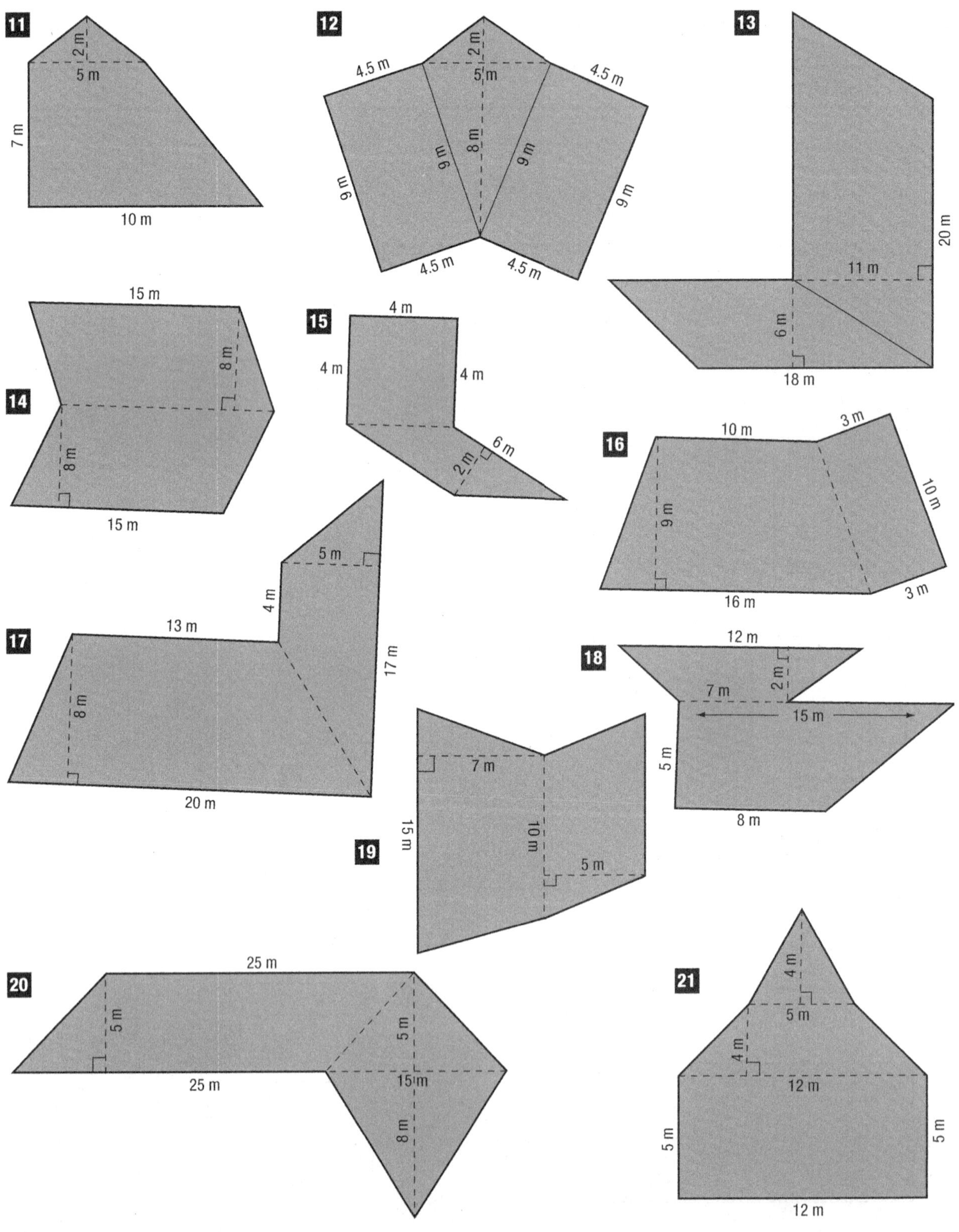
11
2 m
5 m
7 m
10 m
12
4.5 m
2 m
5 m
4.5 m
9 m
8 m
9 m
9 m
9 m
4.5 m
4.5 m
13
20 m
11 m
6 m
18 m
14
15 m
8 m
8 m
15 m
15
4 m
4 m
4 m
6 m
2 m
16
10 m
3 m
9 m
10 m
16 m
3 m
17
5 m
4 m
13 m
17 m
8 m
20 m
18
12 m
2 m
7 m
15 m
5 m
8 m
19
7 m
15 m
10 m
5 m
20
25 m
5 m
5 m
25 m
15 m
8 m
21
4 m
5 m
4 m
12 m
5 m
5 m
12 m

7.2.6 Investigate volumes of compound prismatic solids and use rules to determine volumes

Calculate volume of rectangular prisms

Remember

The formula for finding the volume of a rectangular prism is: length × width × height.

1 Use the given measurements to calculate the volume of each rectangular prism.

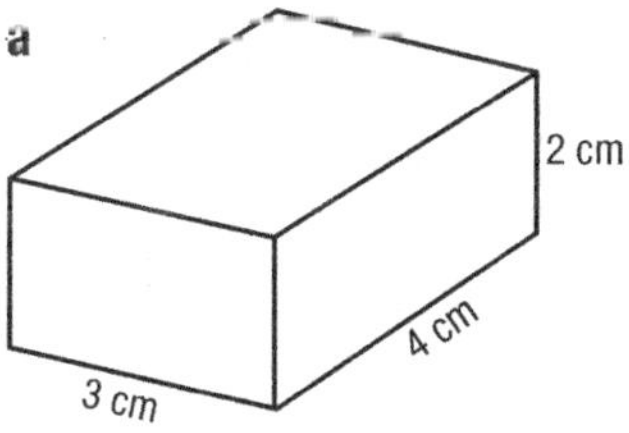

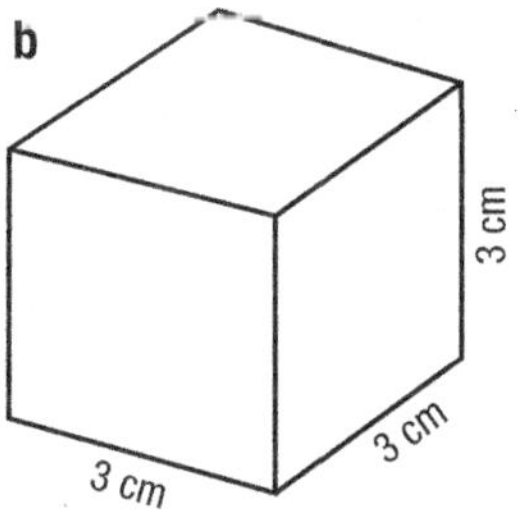

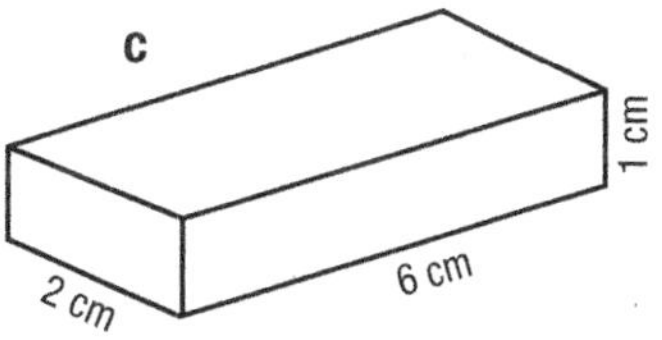

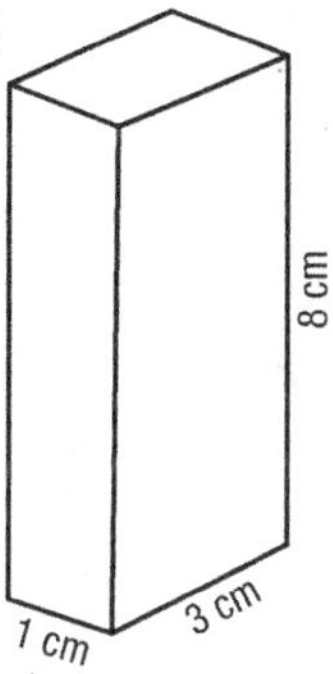

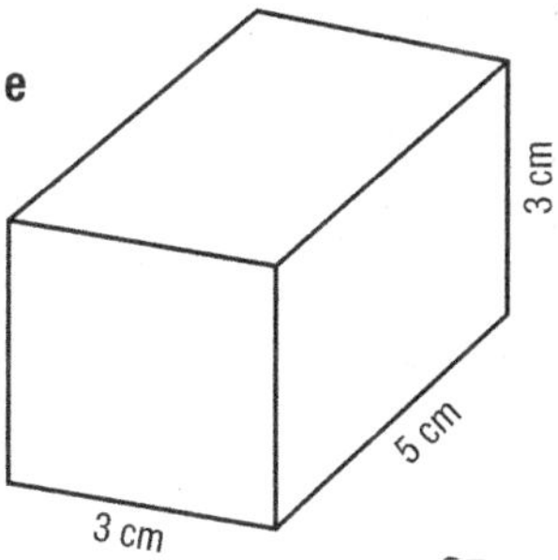

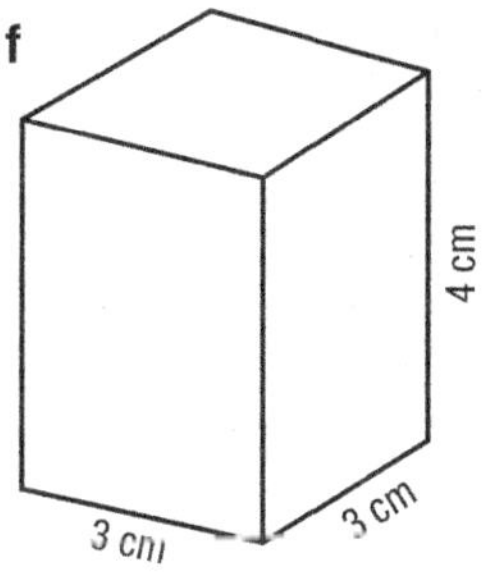

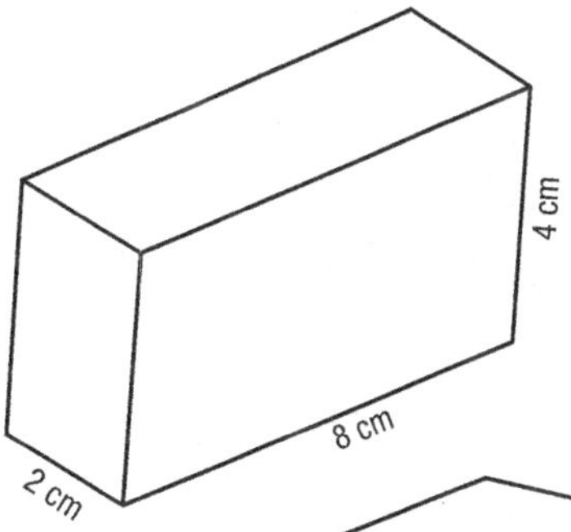

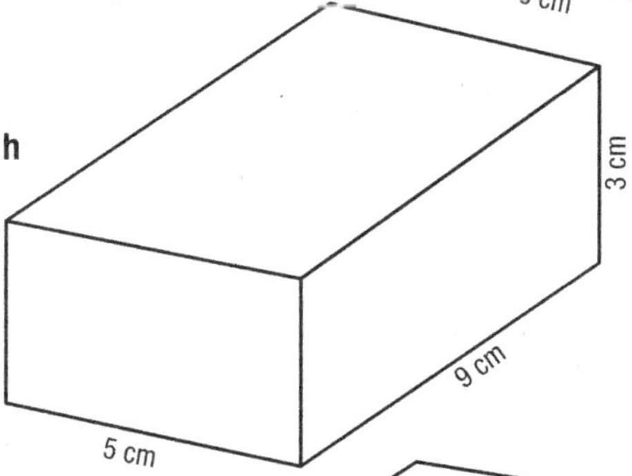

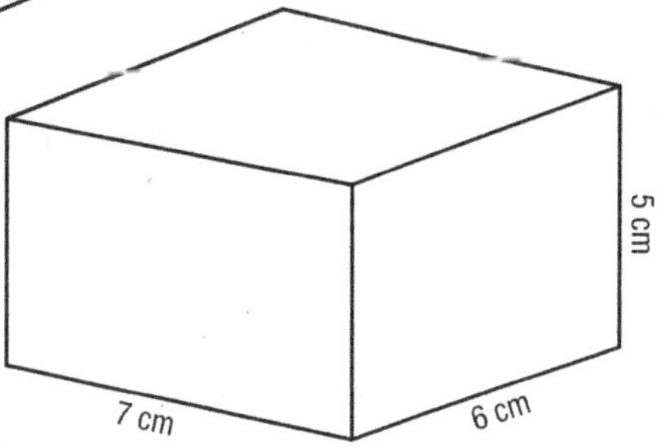

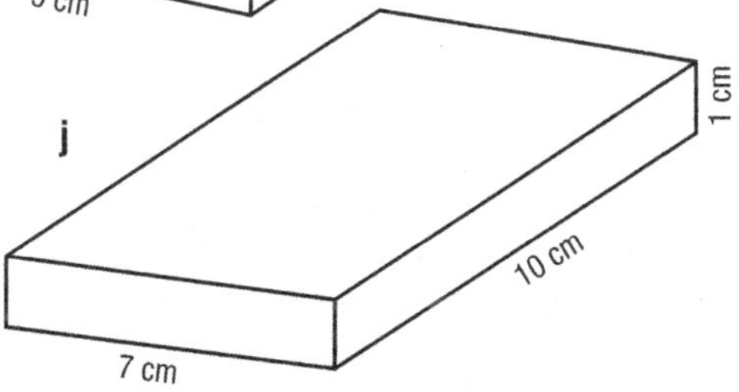

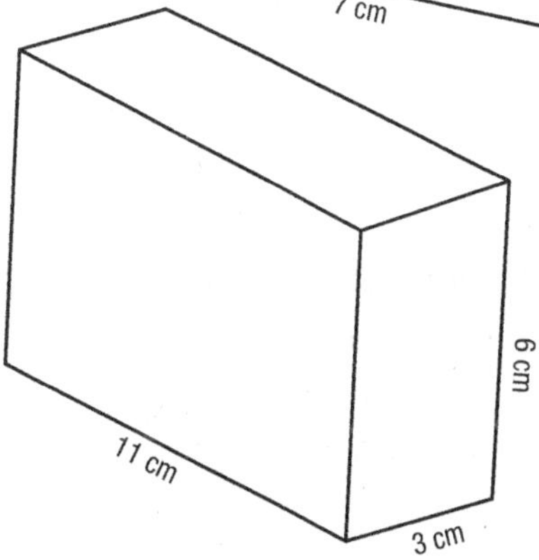

Copy and complete these charts.

2	Length	Width	Height	Volume
a	3 cm	2 cm	5 cm	
b		5 cm	2 cm	100 cm^3
c	6 cm		5 cm	120 cm^3
d	4 cm	2 cm		56 cm^3
e	8 cm	5 cm	9 cm	
f		4 cm	8 cm	128 cm^3
g	7 cm		2 cm	56 cm^3
h	9 cm	3 cm		81 cm^3
i		4 cm	3 cm	144 cm^3
j	5 cm		6 cm	60 cm^3
k	7 cm	7 cm		98 cm^3
l	6 cm		5 cm	90 cm^3
m		9 cm	7 cm	630 cm^3
n	12 cm	6 cm	6 cm	
o	10 cm	11 cm		550 cm^3
p	8 cm	12 cm	3 cm	
q	7 cm	7 cm		343 cm^3
r		9 cm	9 cm	486 cm^3
s	5 cm		8 cm	320 cm^3
t	2 cm	15 cm	11 cm	

3	Length	Width	Height	Volume
a	2.5 cm	4 cm	6.5 cm	
b	8 cm	4.5 cm	2.5 cm	
c	3.5 cm	5 cm	1.2 cm	
d		7 cm	2.5 cm	70 cm^3
e	5 cm	1.8 cm		99 cm^3
f	2.4 cm		10 cm	192 cm^3
g		8 cm	5 cm	180 cm^3
h	9 cm	2.5 cm	9 cm	
i	1.8 cm	5 cm		108 cm^3
j	7 cm		3.6 cm	126 cm^3
k		10 cm	4.3 cm	344 cm^3
l	12 cm	3 cm	5.8 cm	
m	9 cm	9 cm	4.5 cm	
n	2.5 cm	3.5 cm		87.5 cm^3
o	5.6 cm		5 cm	244 cm^3
p		11 cm	5.5 cm	121 cm^3
q	6 cm	8 cm	6.5 cm	
r	1.5 cm	9 cm		162 cm^3
s	15 cm		2.4 cm	144 cm^3
t	8 cm	20 cm	3.5 cm	

4 Calculate the volume of rectangular prisms with the following dimensions.

- **a** 5 cm high, 10 cm long and 3 cm wide
- **b** a cube with 6 m sides
- **c** 8 m long, 4 m high and 5 m wide
- **d** 6 cm wide, 10 cm long and 2 cm high
- **e** a cube with 9 cm sides
- **f** 7 cm high, 8 cm wide and 8 cm long
- **g** 11 cm long, 6 cm wide and 4 cm high
- **h** 3 m wide, 7 m long and 7 m high
- **i** 9 cm high, 11 cm long and 7 cm wide
- **j** a cube with 12 cm sides
- **k** 4 m wide, 6 m high and 15 m long
- **l** 2 cm high, 6 cm long and 4 cm wide
- **m** 9 cm high, 14 cm long and 7 cm wide
- **n** 20 m long, 5 m wide and 12 m high
- **o** a cube with 5 m sides
- **p** 5 cm wide, 6 cm high and 8 cm long
- **q** 12 m long, 4 m wide and 16 m high
- **r** 9 cm long, 7 cm high and 5 cm wide
- **s** 2 cm high, 2 cm wide and 5 cm long
- **t** a cube with 7 cm sides
- **u** 5 cm wide, 11.5 cm long and 2.5 cm high
- **v** 10.5 m long, 8.5 m wide and 6 m high
- **w** 7.5 cm high, 6.2 cm wide and 9 cm long
- **x** 15 m long, 8.4 m wide and 9.5 m high

Calculate volume of triangular prisms

Remember

To find the volume of a triangular prism, first calculate the area of the triangular base and then multiply by the height of the prism.

Use the given measurements to calculate the volume of each triangular prism.

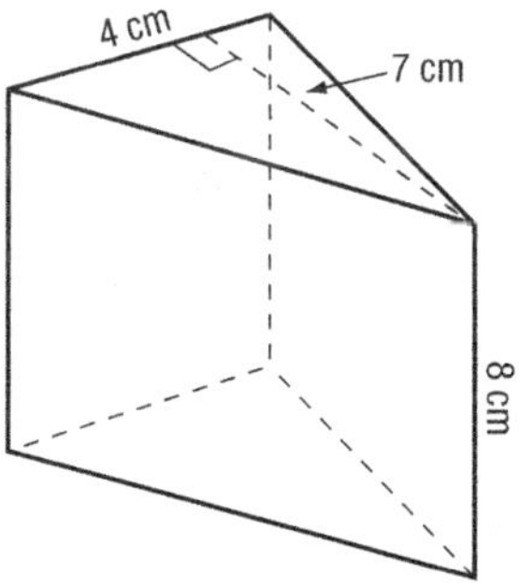

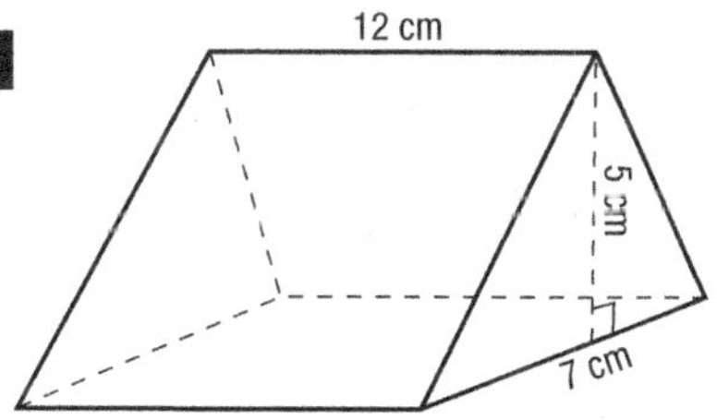

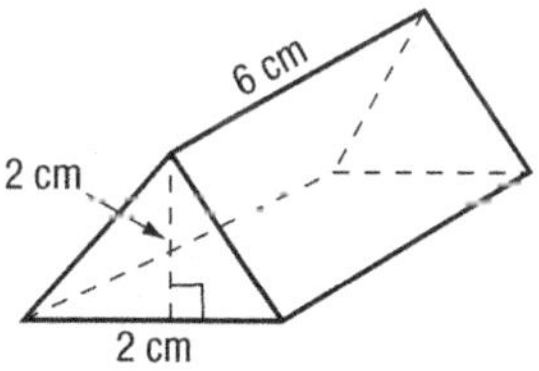

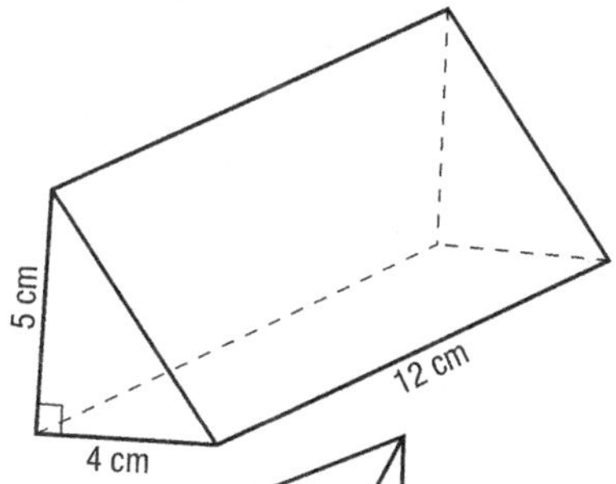

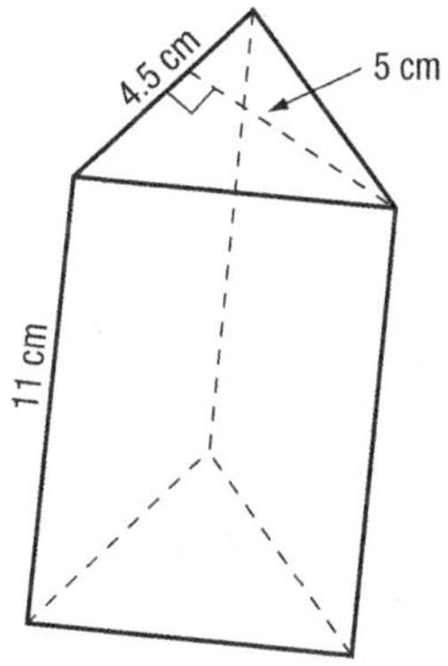

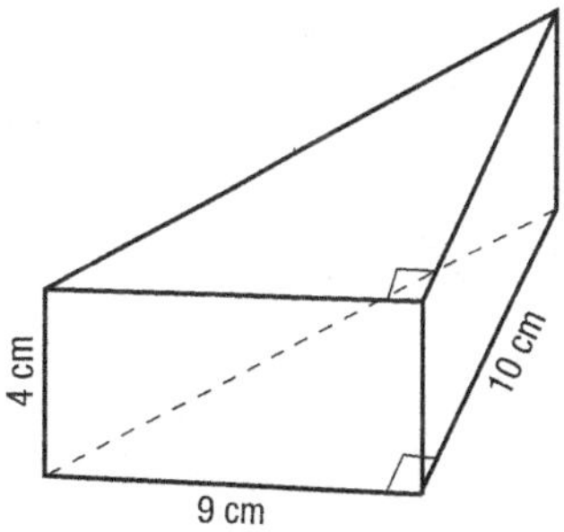

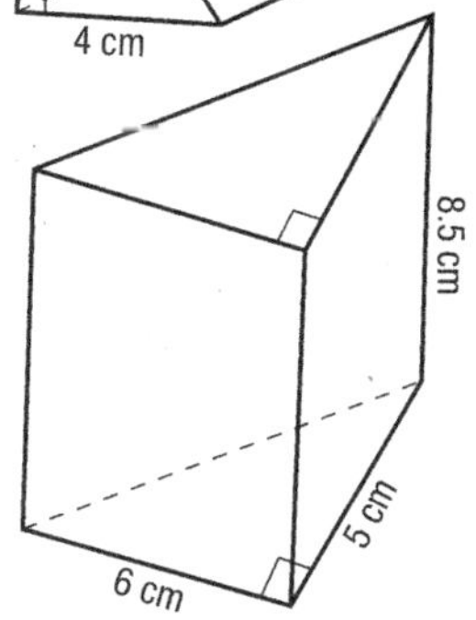

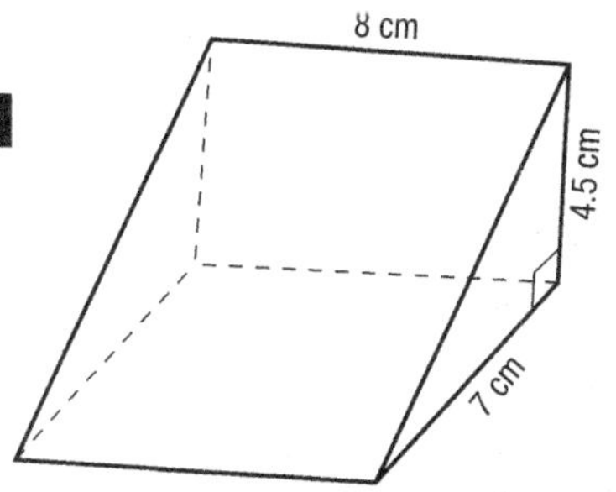

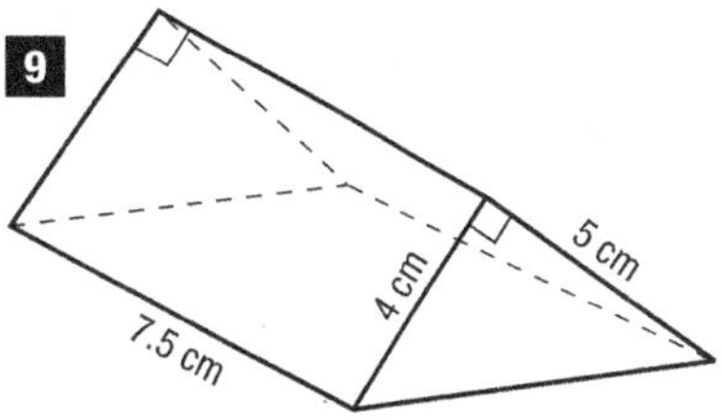

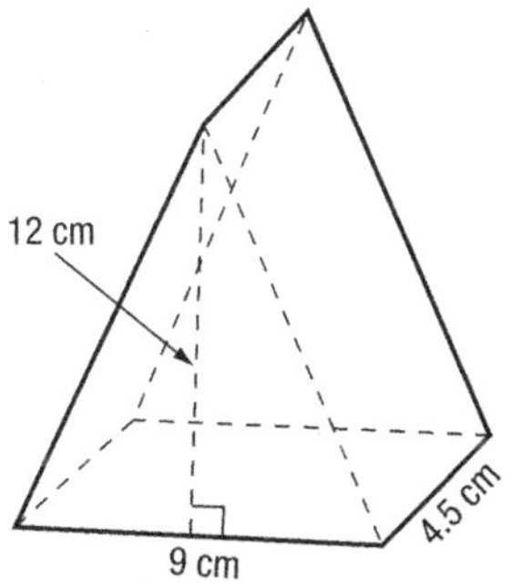

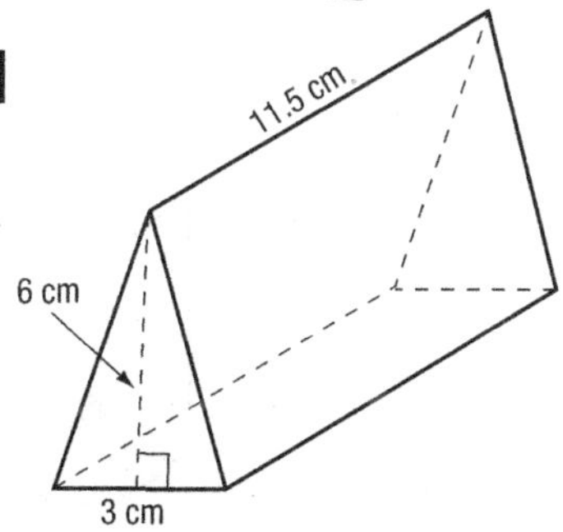

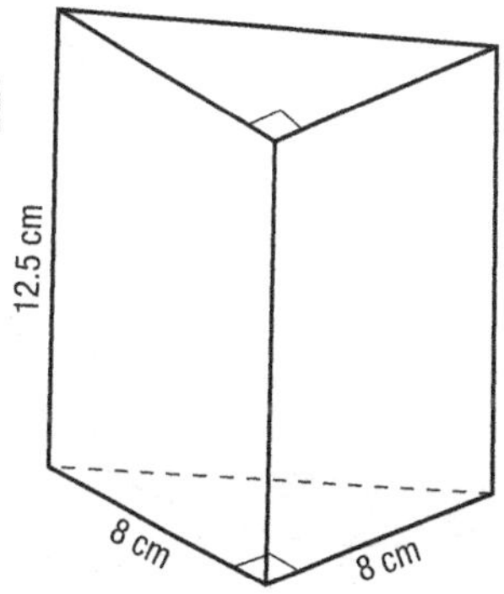

Calculate volume of simple solids

Help Box

If the area of the base of a prism and its height are known, then it is possible to calculate its volume by using the formula: volume = area × height.

Example: the area of the base of the cylinder shown here is 20 cm^2. Its height is 6 cm, so its volume is 20 cm^2 × 6 cm which is 120 cm^3.

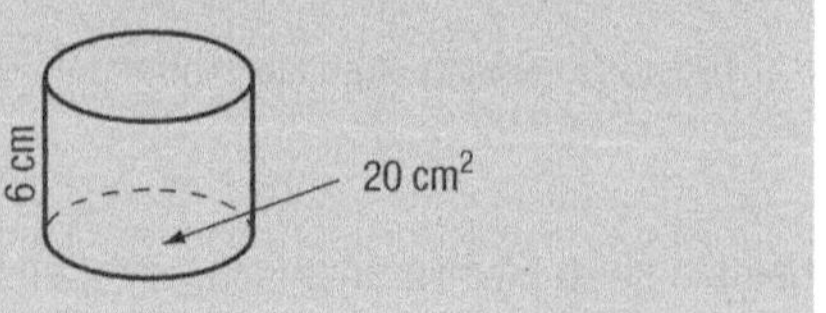

Use the area and height measurements to calculate the volume of each prism.

1 9 cm, 28 cm^2

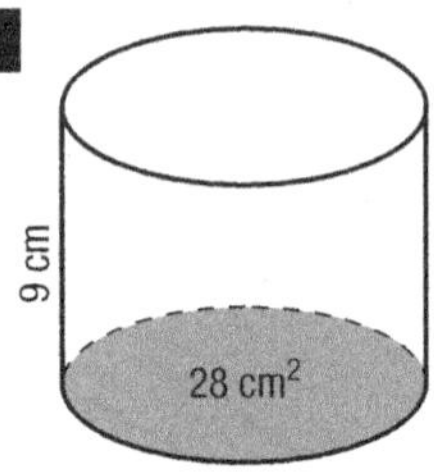

2 15 cm, 21 cm^2

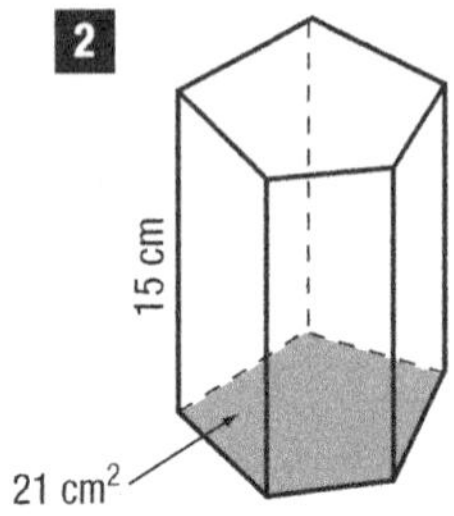

3 7 cm, 36 cm^2

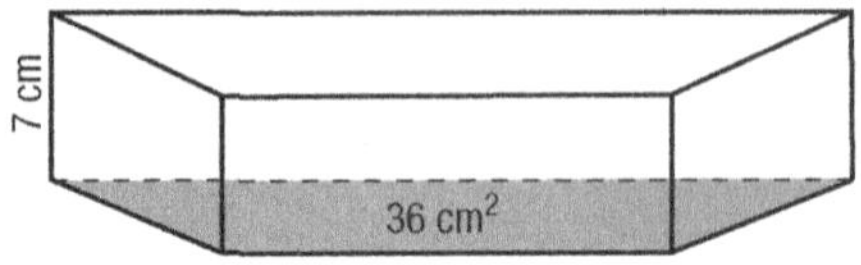

4 25 cm, 20 cm^2

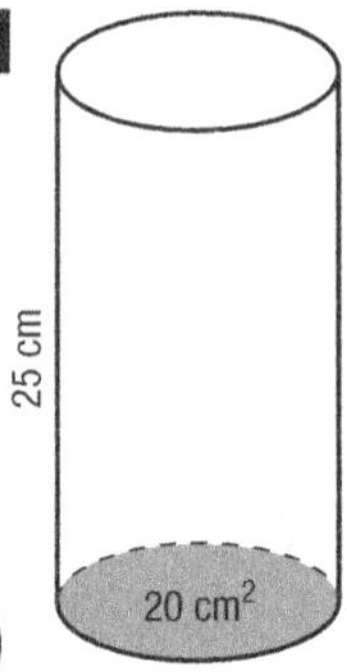

5 10.5 cm, 44 cm^2

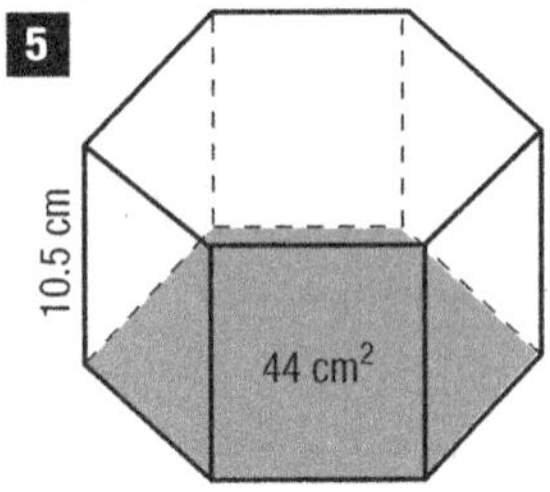

6 11 cm, 39 cm^2

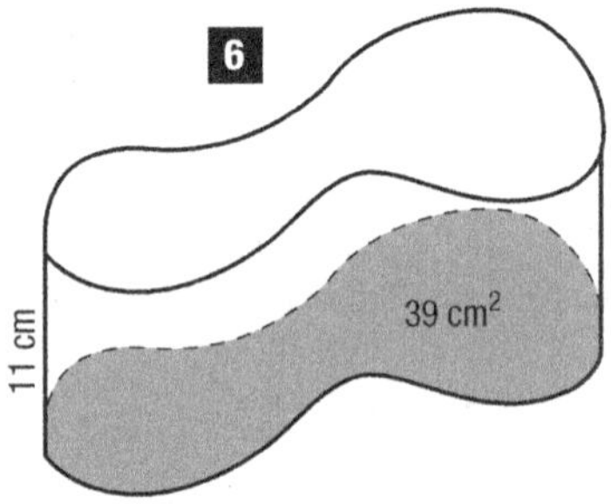

7 17 cm, 52 cm^2

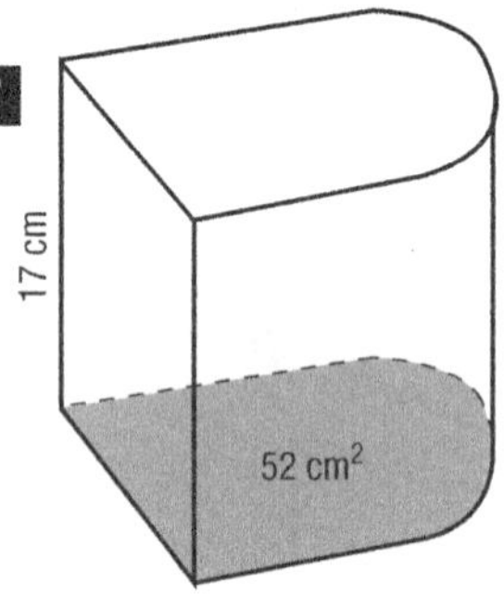

8 12.5 cm, 45 cm^2

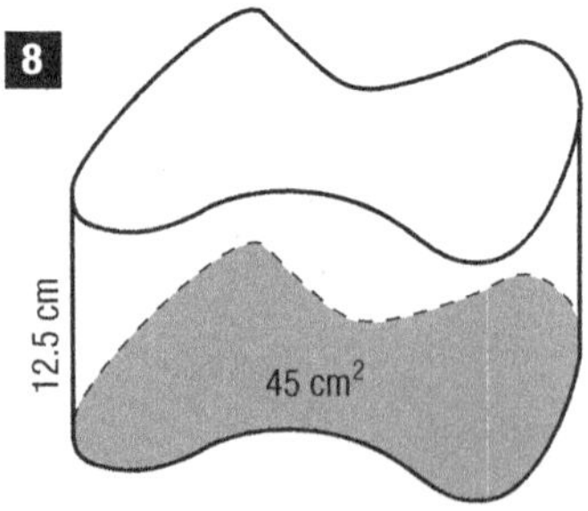

9 18 cm, 42 cm^2

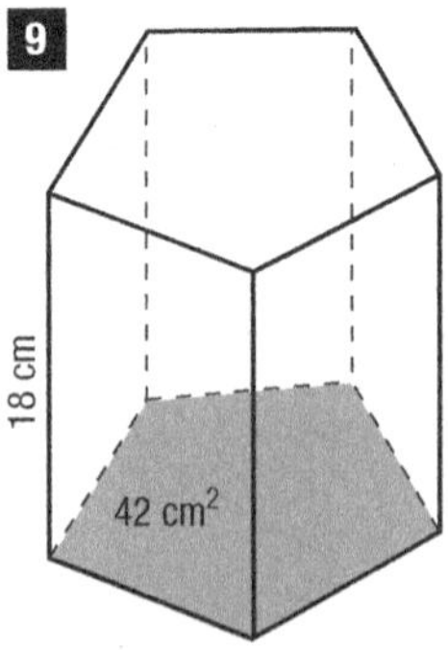

10 22 cm, 55 cm^2

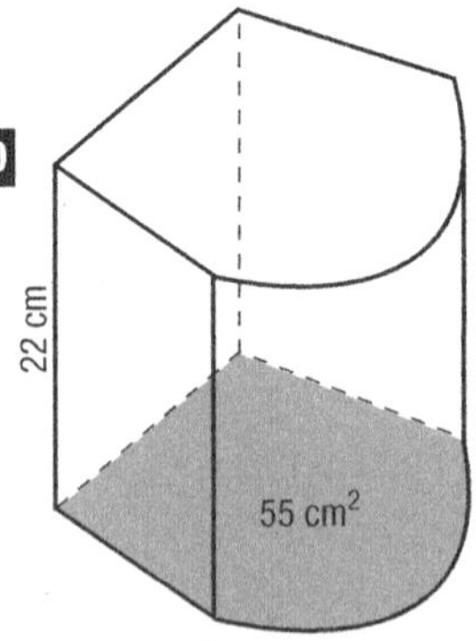

11 15 cm, 32 cm^2

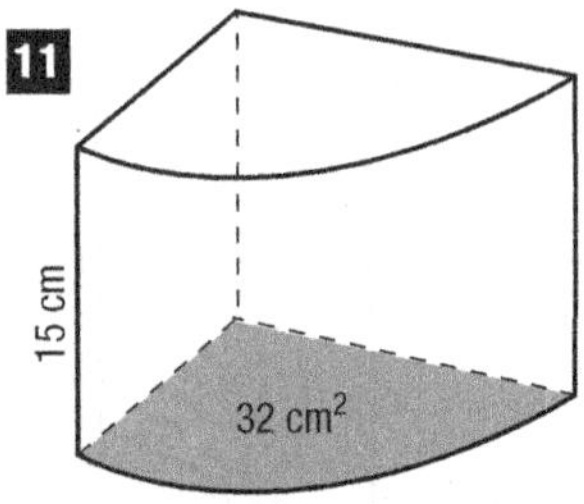

12 16.5 cm, 38 cm^2

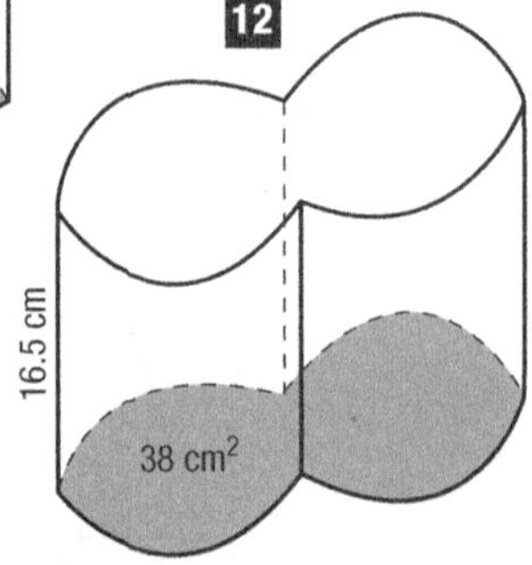

Calculate volume of compound solids

Remember

When solids are made up of a combination of smaller solids, find the volume of each of the smaller solids and then add them together to find the volume of the compound solid.

Use the given measurements to calculate the volume of each shape.

1

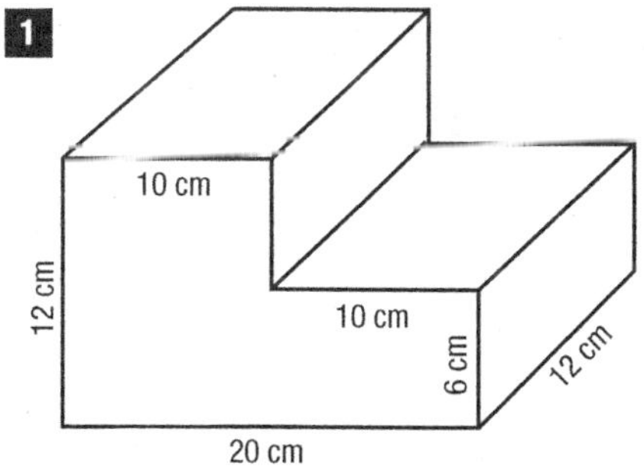

10 cm, 12 cm, 10 cm, 6 cm, 12 cm, 20 cm

2

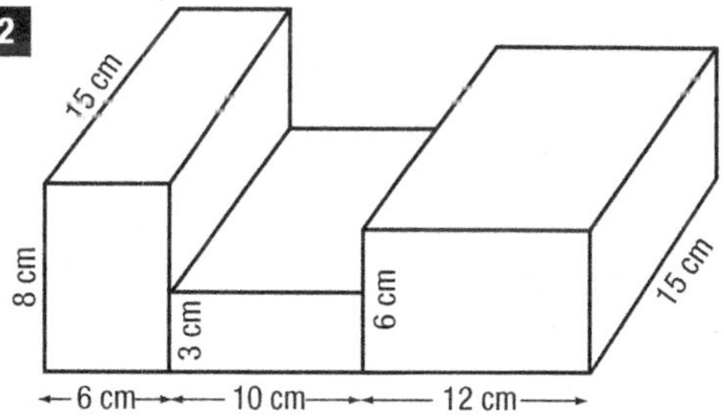

15 cm, 8 cm, 3 cm, 6 cm, 15 cm, 6 cm, 10 cm, 12 cm

3

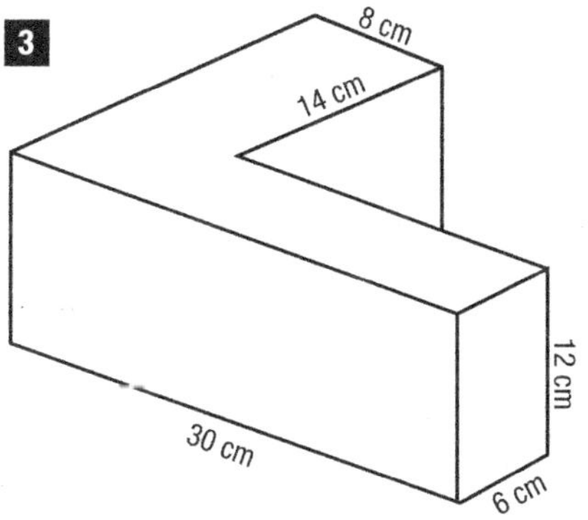

8 cm, 14 cm, 12 cm, 30 cm, 6 cm

4

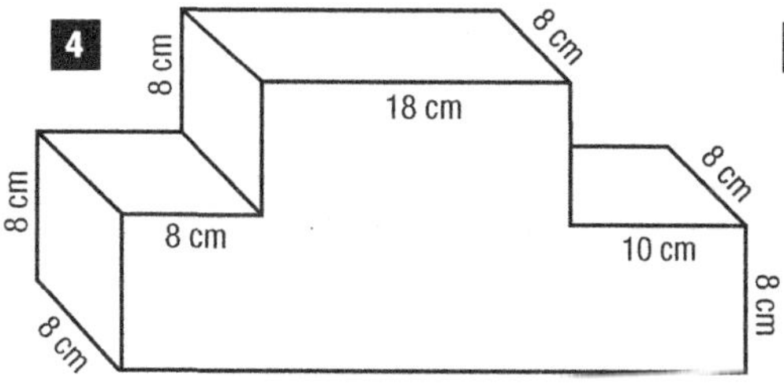

8 cm, 8 cm, 18 cm, 8 cm, 8 cm, 8 cm, 10 cm, 8 cm, 8 cm

5

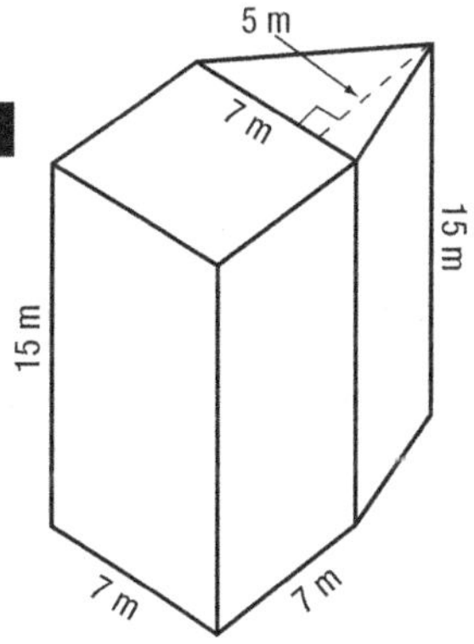

5 m, 7 m, 15 m, 15 m, 7 m, 7 m

6

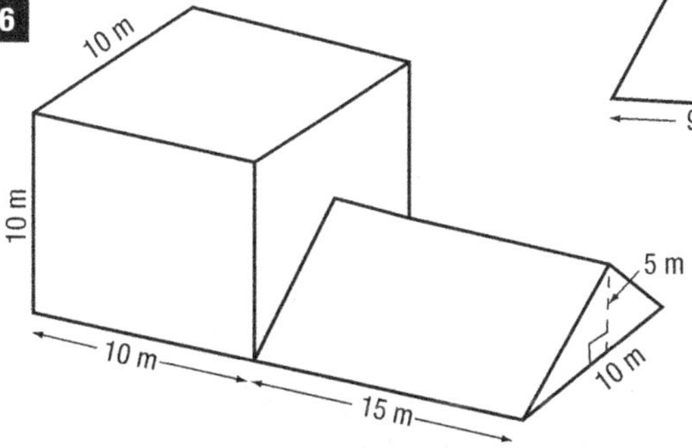

10 m, 10 m, 5 m, 10 m, 10 m, 15 m

7

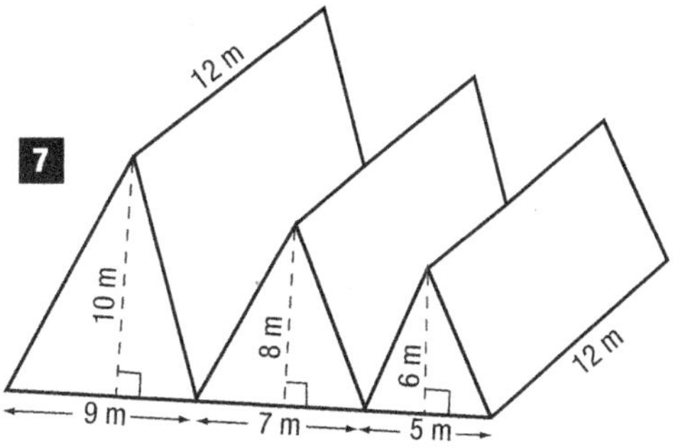

12 m, 10 m, 8 m, 6 m, 12 m, 9 m, 7 m, 5 m

8

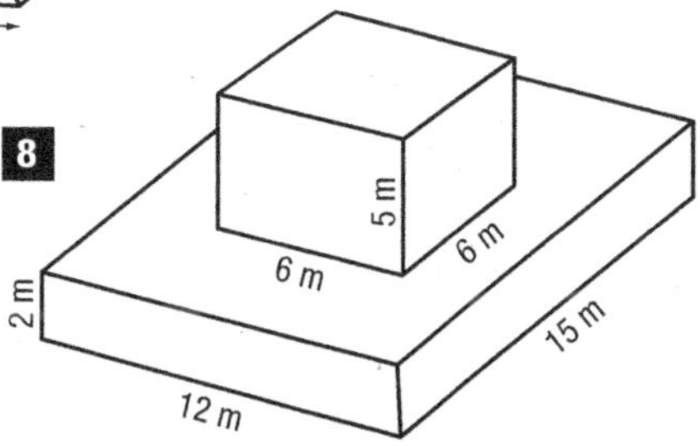

5 m, 6 m, 6 m, 2 m, 15 m, 12 m

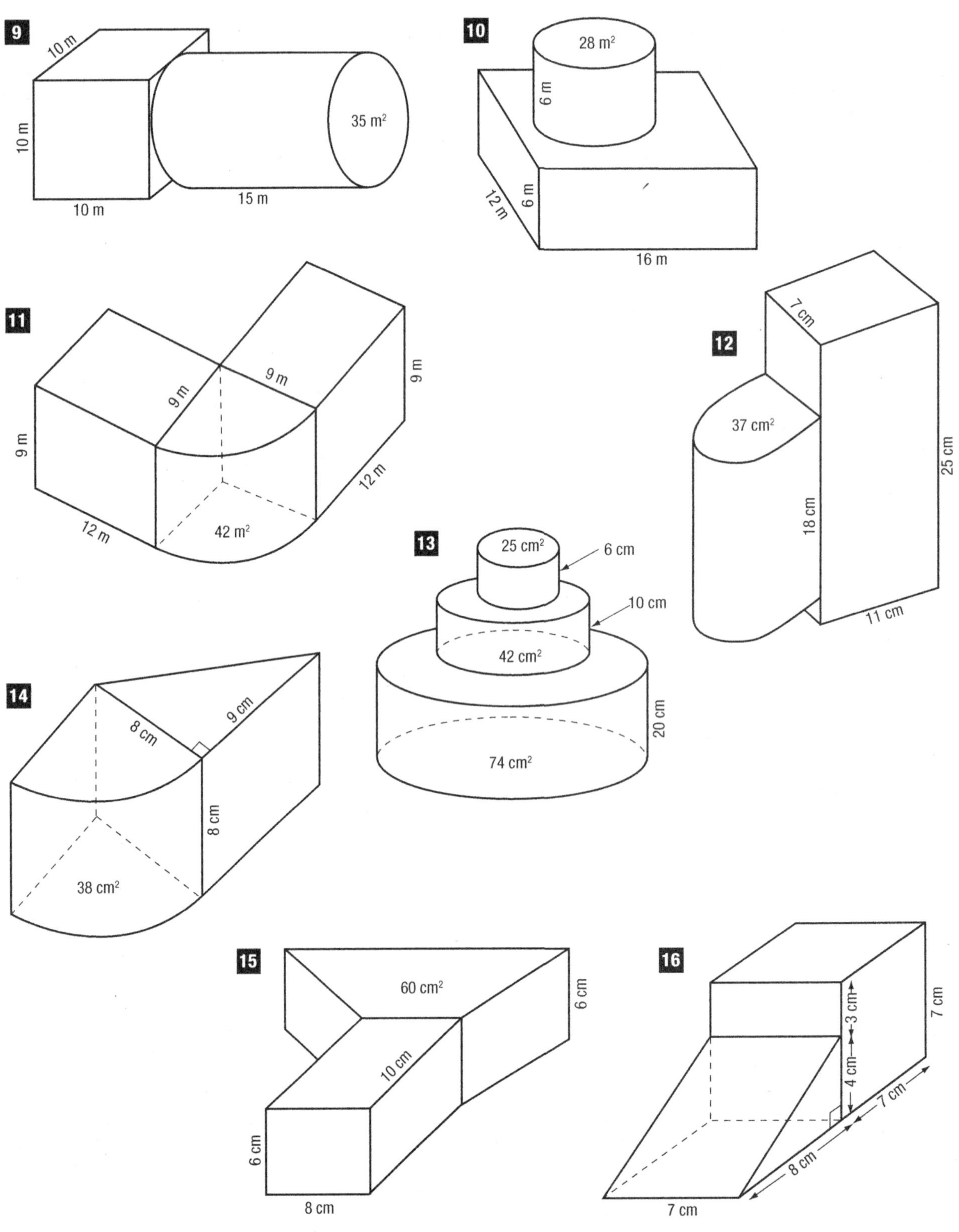
9
10 m
10 m
10 m
35 m²
15 m
10
28 m²
6 m
12 m
6 m
16 m
11
9 m
9 m
9 m
9 m
12 m
12 m
42 m²
12
7 cm
37 cm²
18 cm
25 cm
11 cm
13
25 cm²
6 cm
10 cm
42 cm²
20 cm
74 cm²
14
8 cm
9 cm
8 cm
38 cm²
15
60 cm²
6 cm
10 cm
6 cm
8 cm
16
3 cm
7 cm
4 cm
7 cm
8 cm
7 cm

7.2.7 Investigate the relationship between capacity and volume

Relate capacity and volume

Remember

There is a connection between capacity and volume. 1 mL = 1 cm^3, 1 L = 1000 cm^3 and 1000 L = 1 m^3.

1 What is the volume of containers with the following capacity?

a	10 mL	b	250 mL	c	3 L	d	700 mL	e	5 L	f	1.5 L
g	475 mL	h	2.25 L	i	85 mL	j	830 mL	k	1.75 L	l	9.5 L
m	3.4 L	n	12 L	o	374 mL	p	15 L				

2 What is the capacity of containers with the following volume?

a	20 cm^3	b	65 cm^3	c	1200 cm^3	d	250 cm^3	e	4000 cm^3
f	1600 cm^3	g	715 cm^3	h	535 cm^3	i	7500 cm^3	j	325 cm^3
k	20 000 cm^3	l	17 000 cm^3	m	116 cm^3	n	8300 cm^3	o	14 500 cm^3

3 When an object is put into a container of liquid, it takes up its volume in liquid and causes the liquid to rise in the container. What is the volume of objects that displace the following amounts of liquid?

a	230 mL	b	180 mL	c	6 L	d	95 mL	e	710 mL
f	9 L	g	5.5 L	h	1360 mL	i	538 mL	j	2.75 L
k	4.1 L	l	755 mL	m	6.7 L	n	107 mL	o	1146 mL

4 How much liquid would objects of the following volumes displace?

a	650 cm^3	b	700 cm^3	c	1000 cm^3	d	3000 cm^3	e	285 cm^3
f	3500 cm^3	g	410 cm^3	h	45 cm^3	i	90 cm^3	j	935 cm^3
k	7600 cm^3	l	1250 cm^3	m	4400 cm^3	n	11 000 cm^3	o	13 500 cm^3

5 What is the volume of containers with the following capacity?

a	2000 L	b	5000 L	c	1500 L	d	8500 L	e	3500 L
f	6250 L	g	2750 L	h	10 000 L	i	15 000 L	j	27 000 L
k	18 500 L	l	16 500 L	m	12 250 L	n	33 250 L	o	13 750 L

6 What is the capacity of containers with the following volume?

a	3 m^3	b	8 m^3	c	5 m^3	d	2.5 m^3	e	4.5 m^3
f	10 m^3	g	11.5 m^3	h	14 m^3	i	7.25 m^3	j	10.25 m^3
k	9.75 m^3	l	15.75 m^3	m	20 m^3	n	35 m^3	o	26 m^3

7.2.8 Use appropriate units for capacity and solve capacity problems

Use volume to determine capacity

Remember

Once the volume of a container is known, its capacity can be found by converting the cubic measurement to litres or millilitres.

Find the capacity of containers with these dimensions.

1

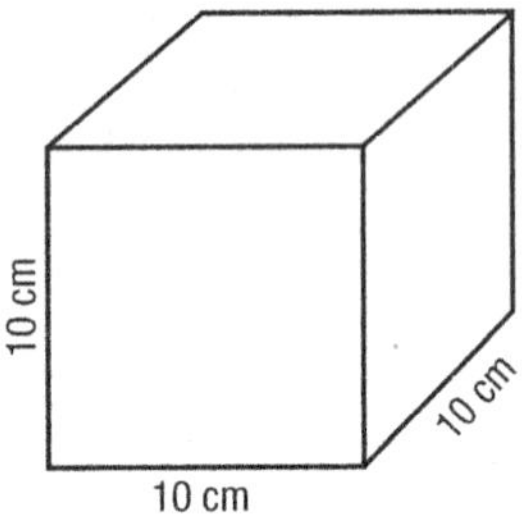

2

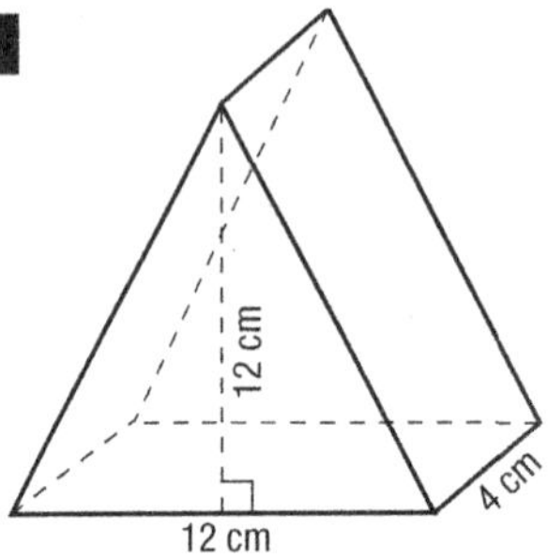

3

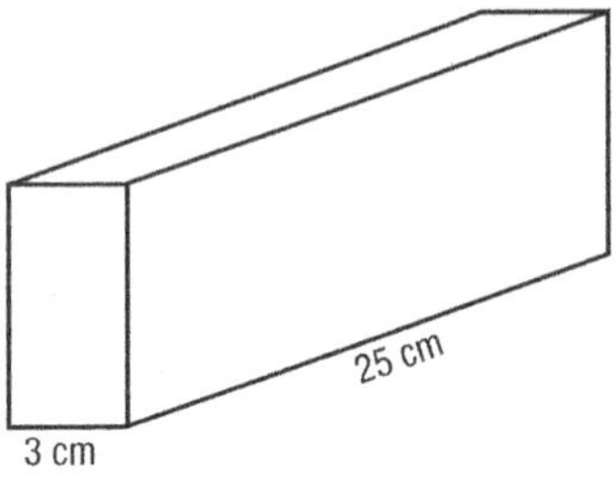

4

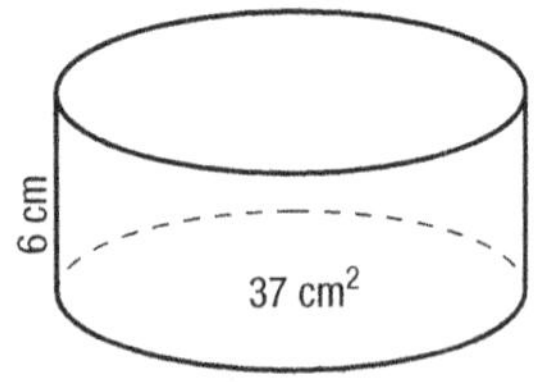

5

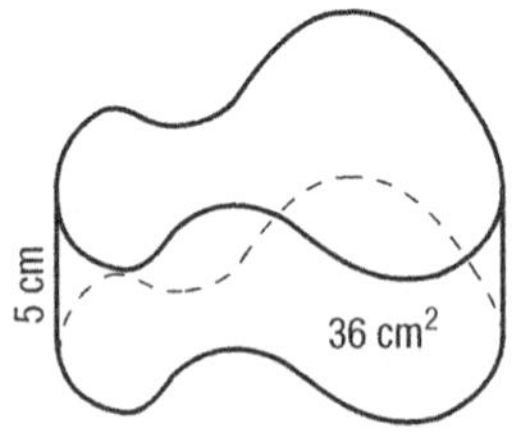

6

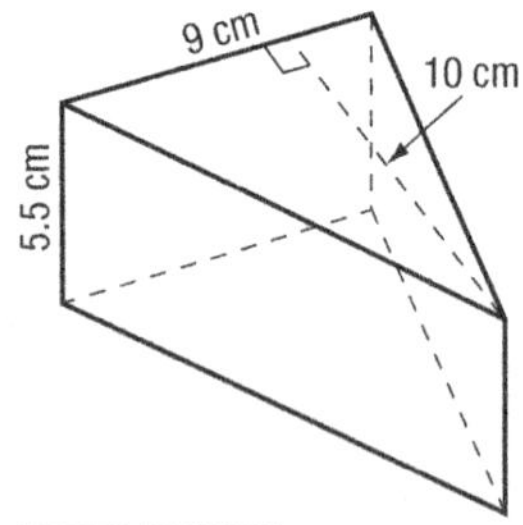

7

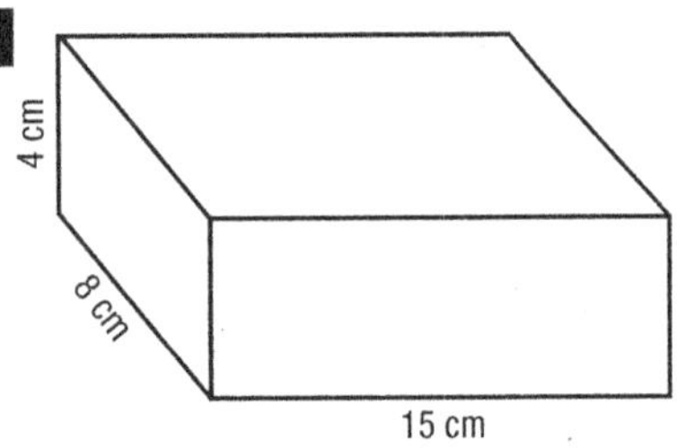

8

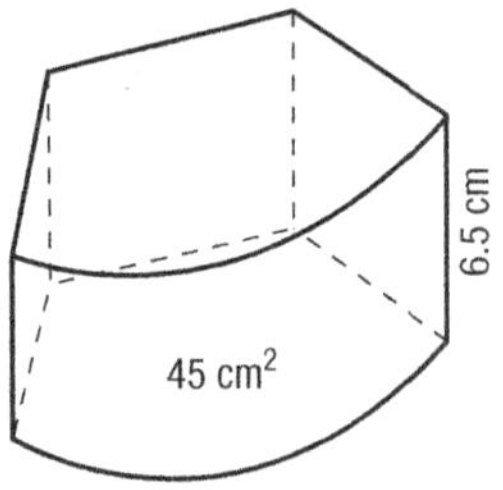

9

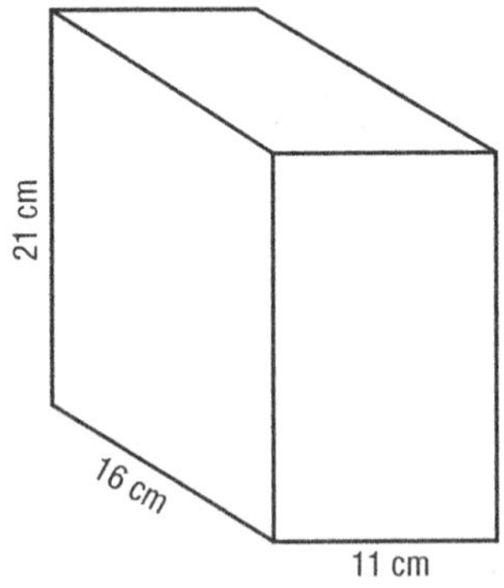

10

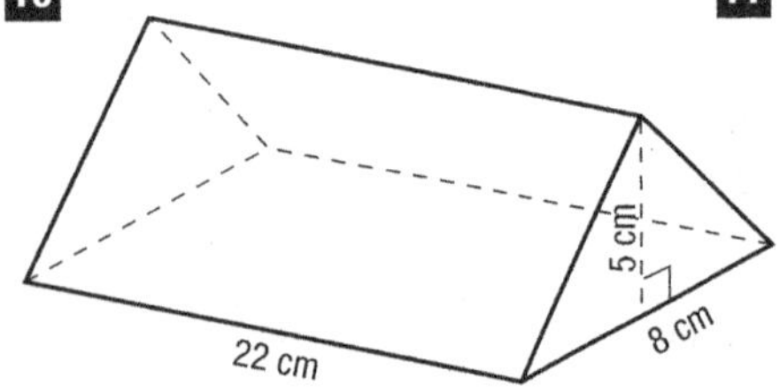

11

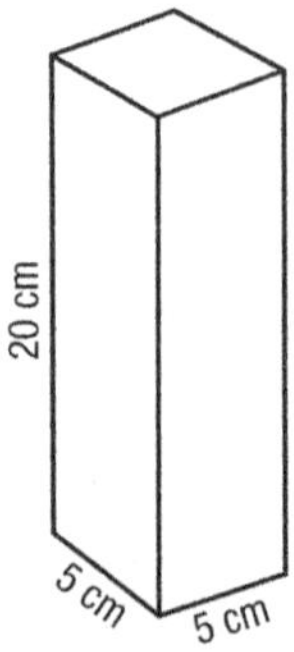

12

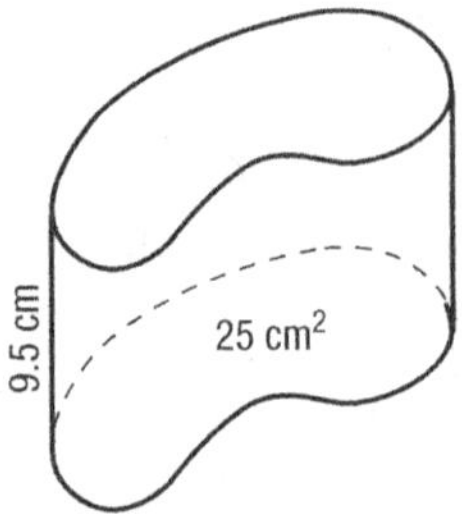

7.2.9 Draw, investigate and make physical models of quadrilaterals

Identify properties of quadrilaterals

Remember

Rectangles, squares, rhombuses and parallelograms are all part of the family of parallelograms. Squares also belong to the family of rhombuses and the family of kites.

Which of the following shapes are:

1 parallelograms? **2** rhombuses? **3** trapeziums? **4** kites?

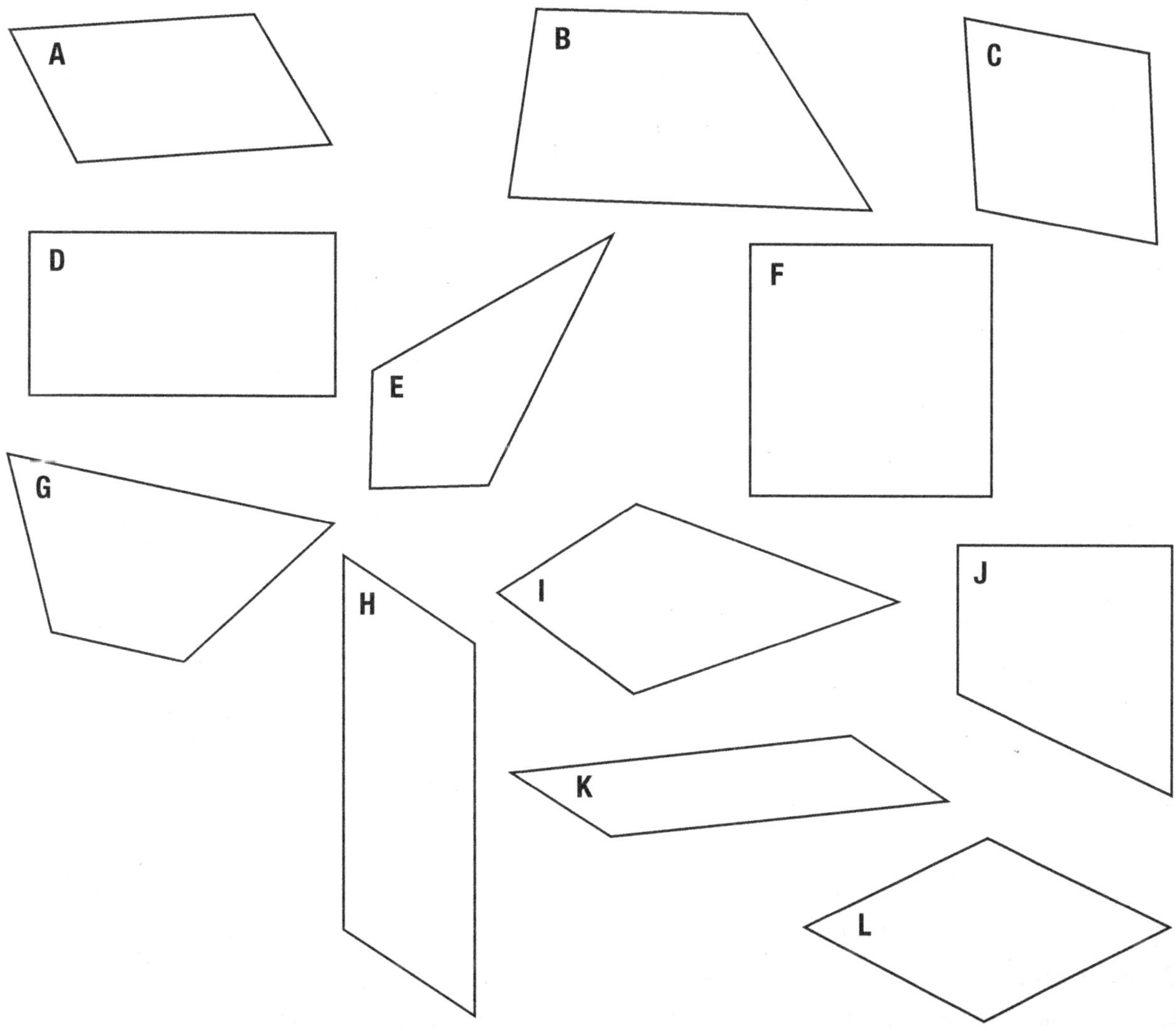

A B C D E F G H I J K L

Classify parallelograms

Remember

Rectangles, squares, rhombuses and parallelograms are all part of the family of parallelograms.

1 Draw a chart like this one, but with enough rows to fit all the shapes, and complete it by looking at each shape below.

Shape	Name of shape	Number of axes (lines of symmetry)
A		
B		
C		

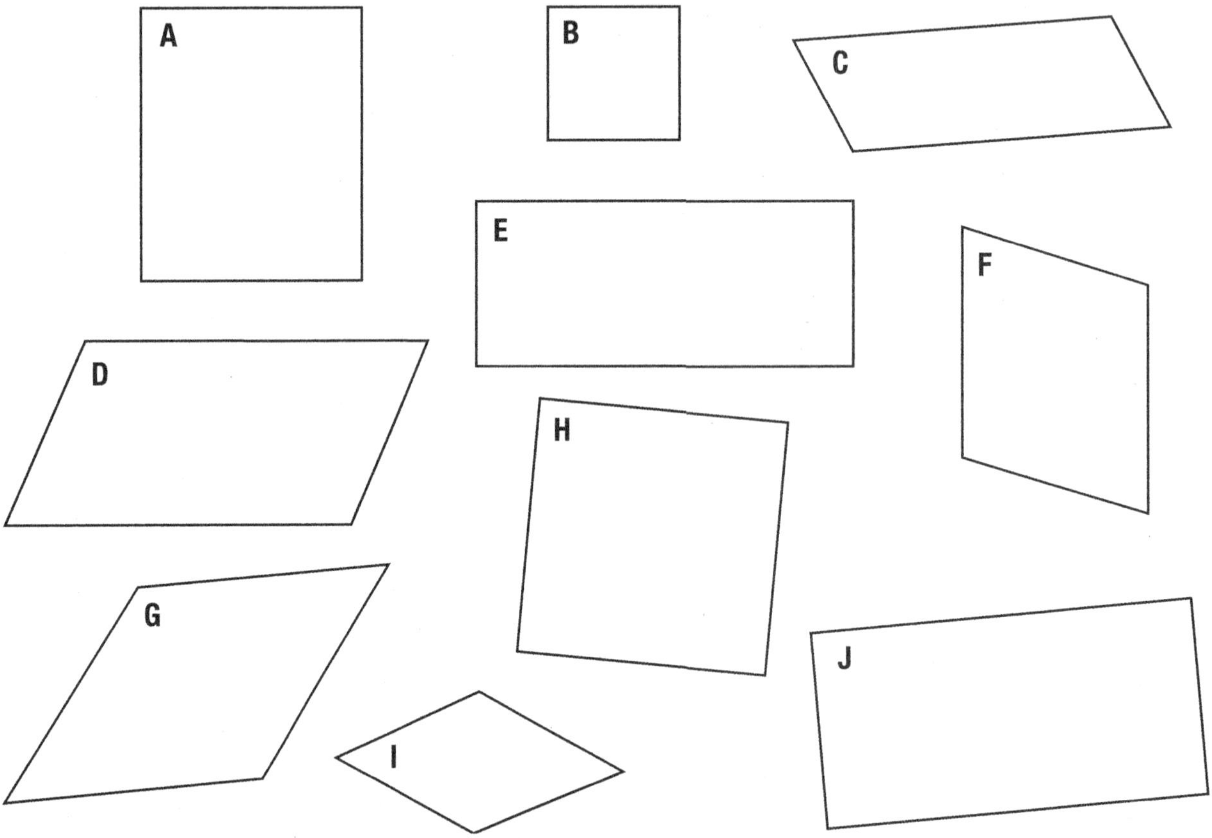

A

B

C

E

F

D

H

G

J

I

2 How many axes of symmetry do rectangles have?

3 How many axes of symmetry do squares have?

4 How many axes of symmetry do rhombuses that are not also squares have?

5 How many axes of symmetry do parallelograms that are not also squares, rectangles or rhombuses have?

Determine parallelograms

Remember

The opposite angles of parallelograms are equal.

Measure the opposite angles in each quadrilateral to find the parallelograms.

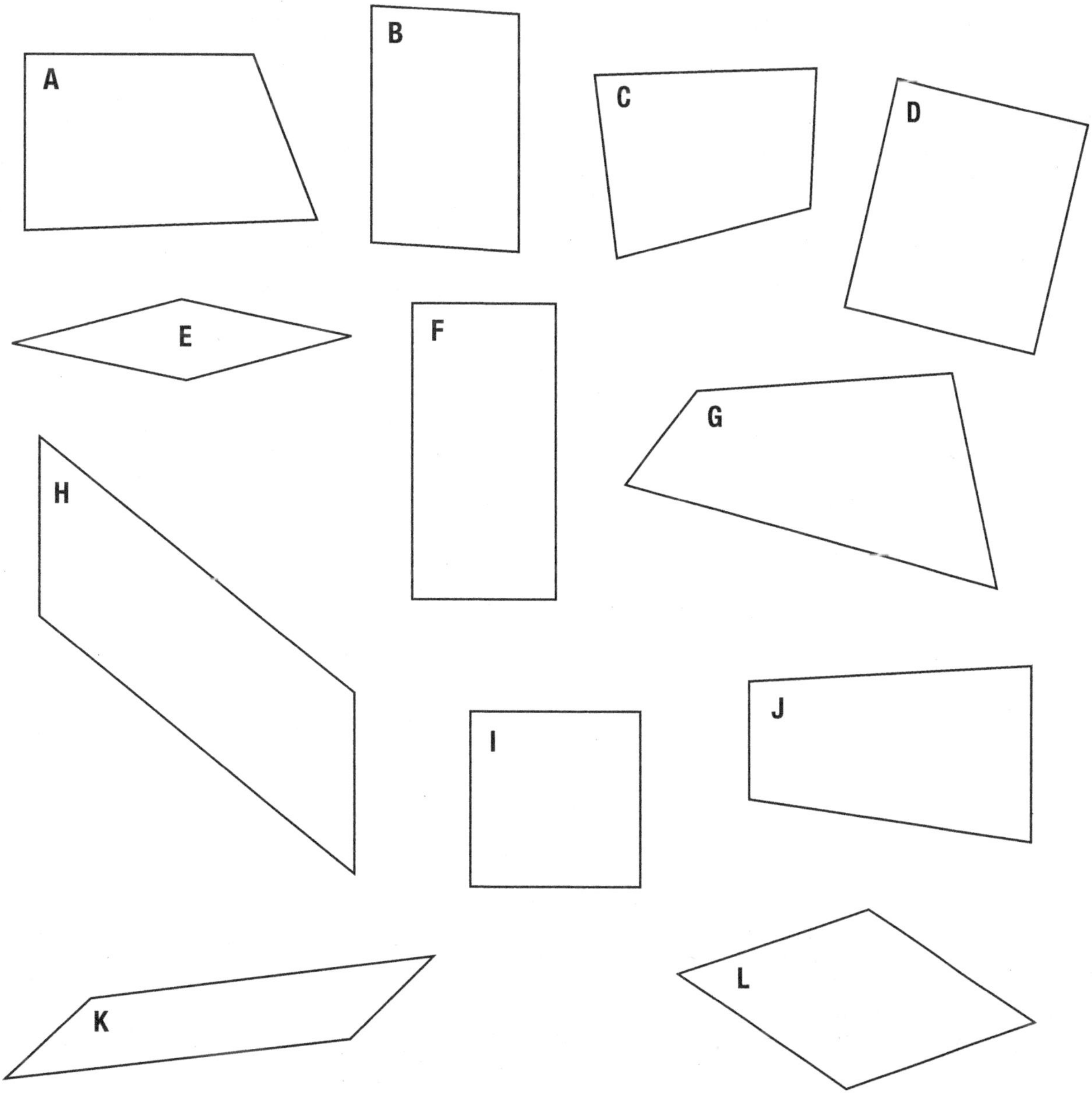

A

B

C

D

E

F

G

H

I

J

K

L

7.2.11 Determine the interior and exterior angles of triangles and quadrilaterals

Calculate interior angles in triangles

Remember

The interior angles of a triangle add up to 180°.

Use what you know about the interior angle sum of triangles to calculate the unknown angle in each triangle.

1 75°, 65°

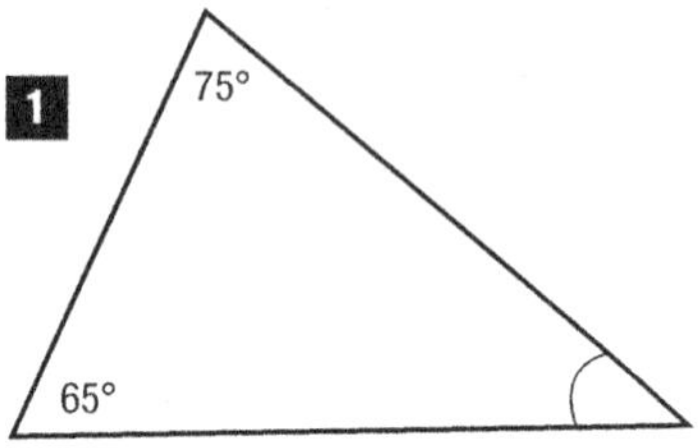

2 60°, 60°

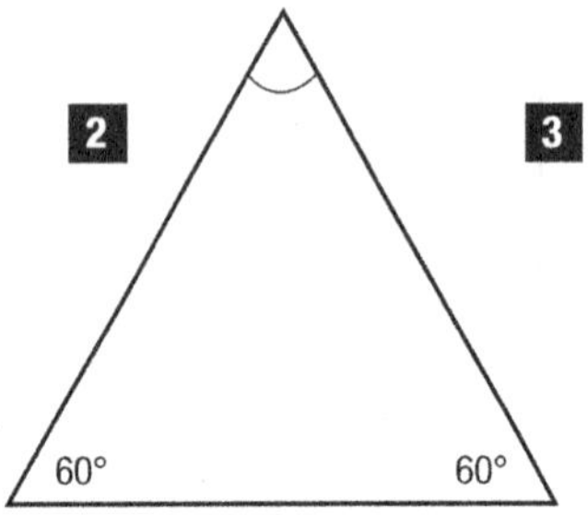

3 100°, 45°

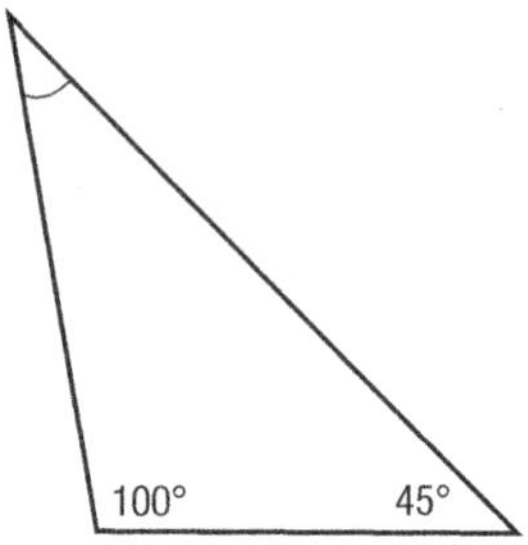

4 75°, 35°

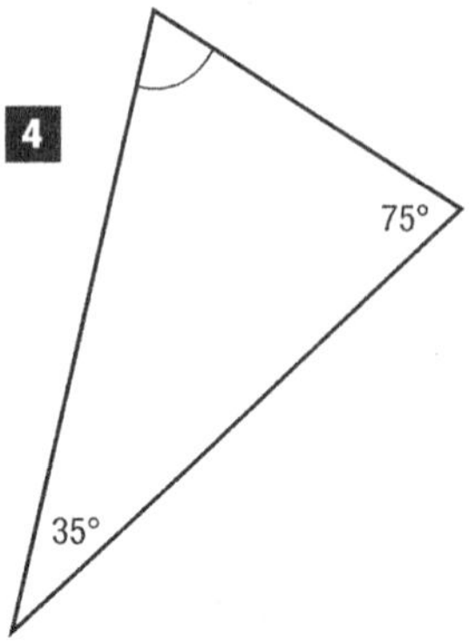

5 65°, 60°

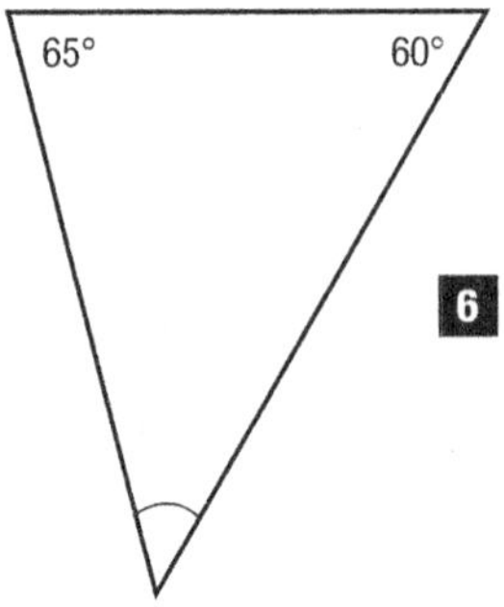

6 70°, 55°

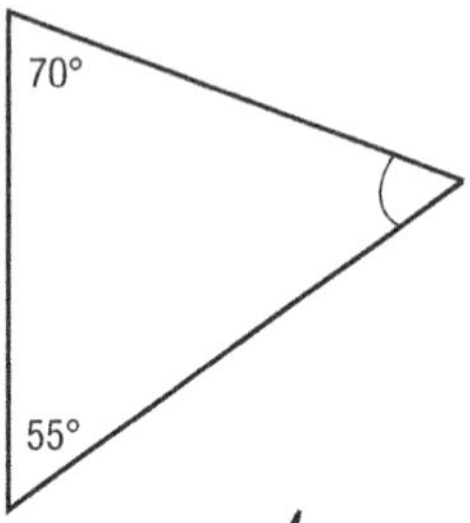

7 110°, 35°

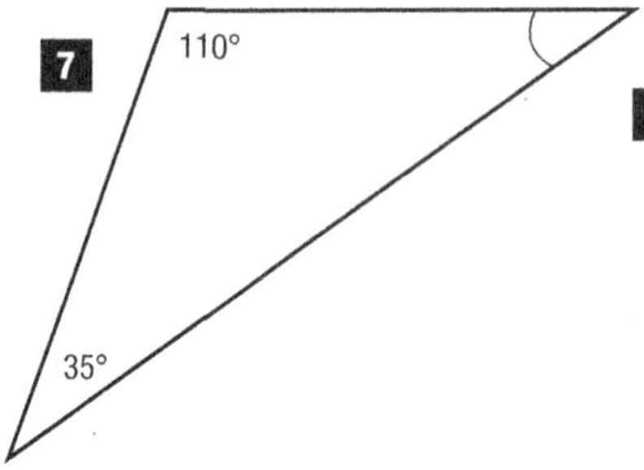

8 50°, 105°

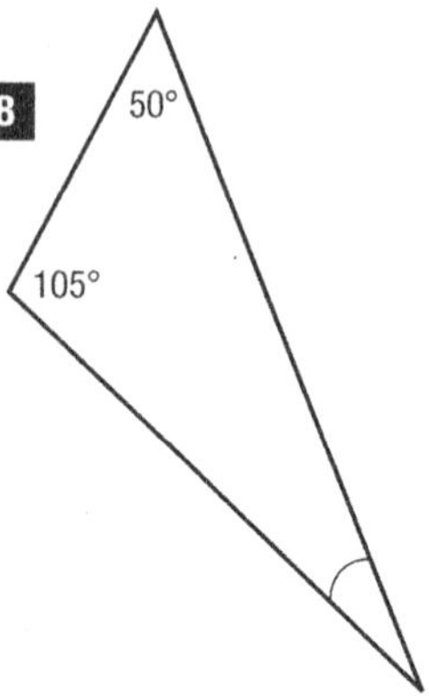

9 65°, 90°

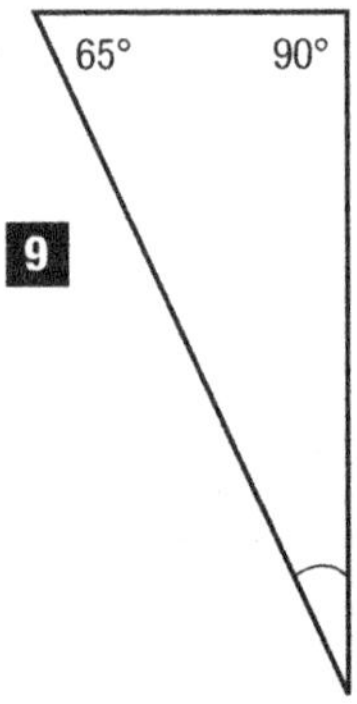

10 15°, 60°

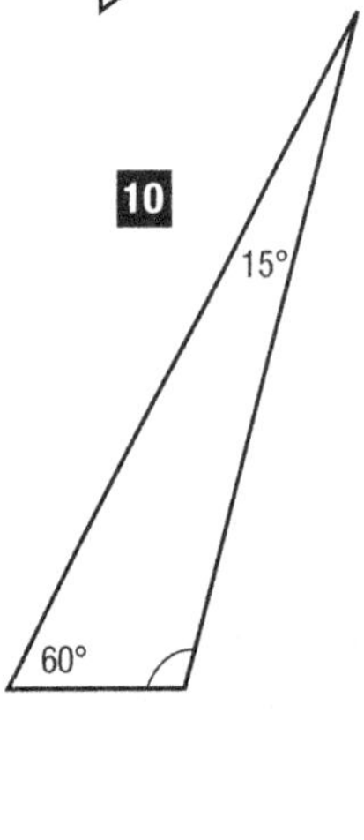

11 115°, 40°

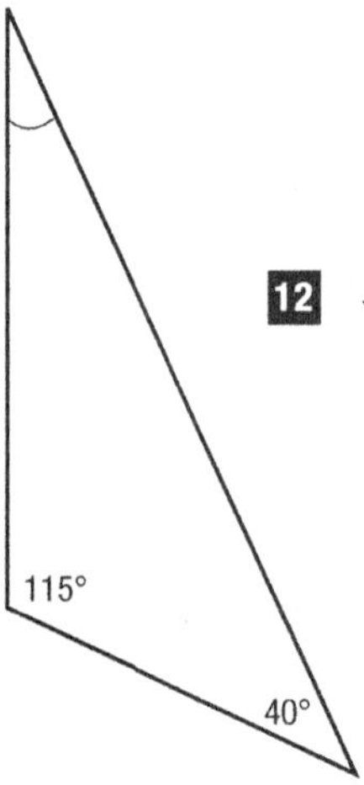

12 40°, 80°

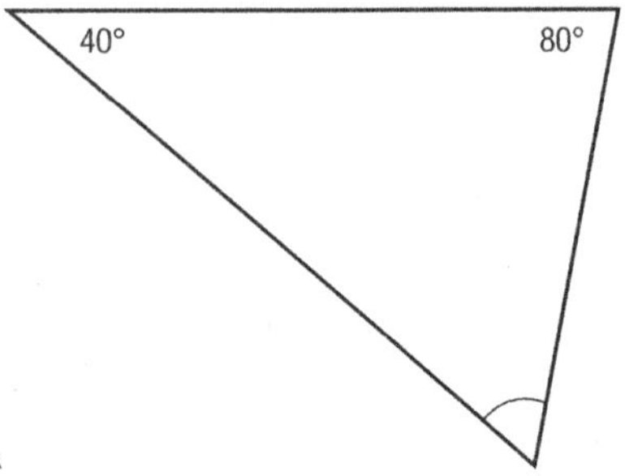

Calculate interior angles in quadrilaterals

Remember

The interior angles of a quadrilateral add up to 360°.

Use what you know about the interior angle sum of quadrilaterals to calculate the unknown angle in each quadrilateral.

1

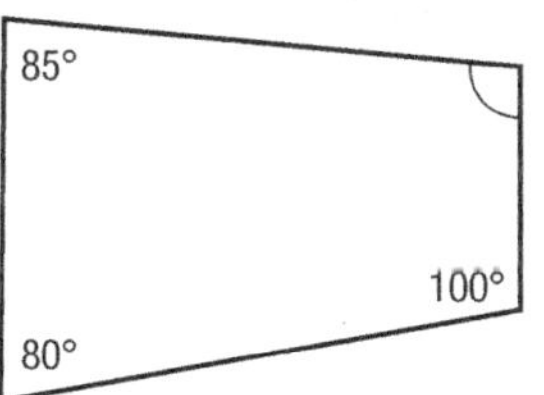

2

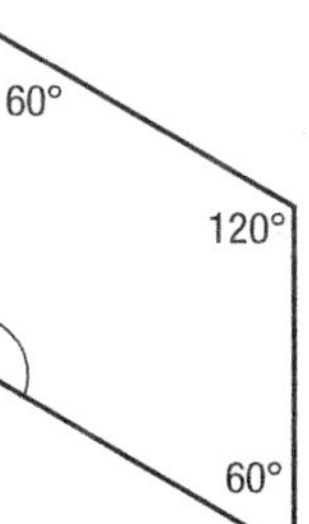

3

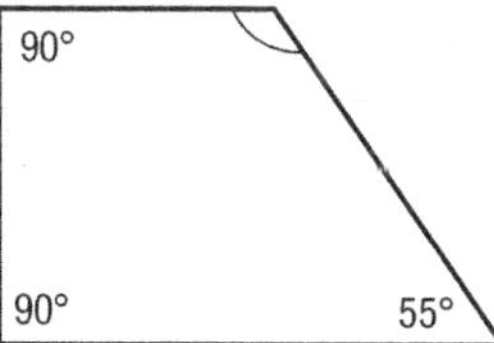

4

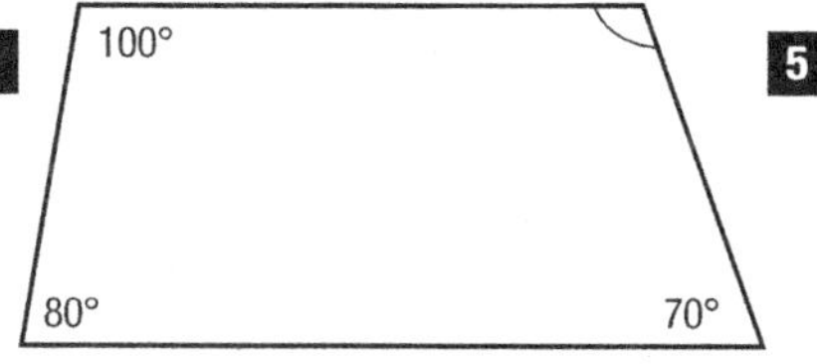

5

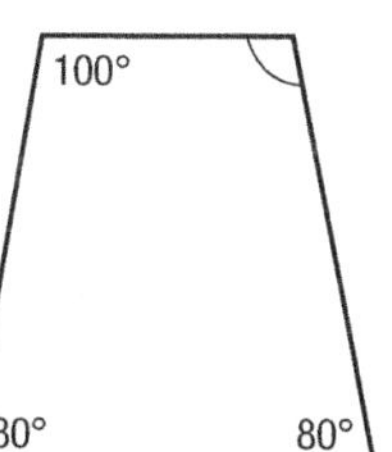

6

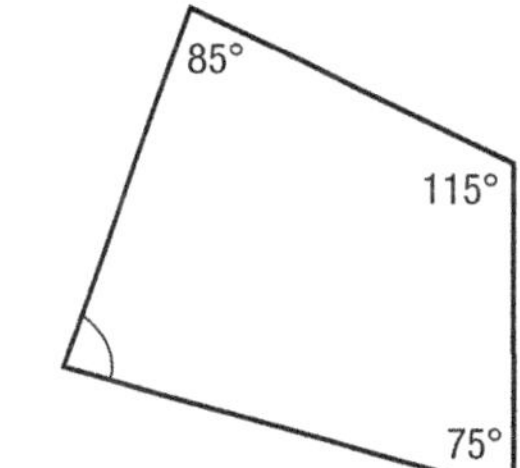

7

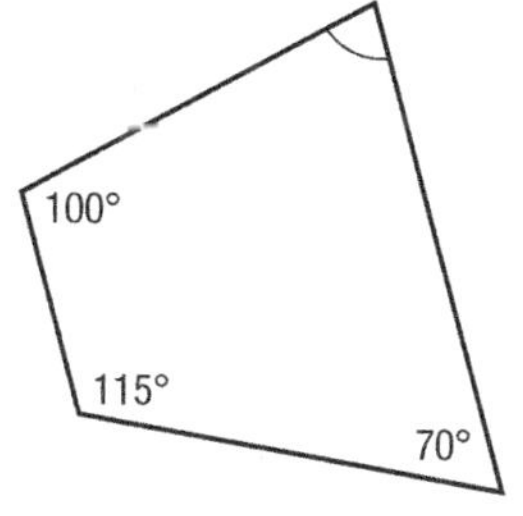

8

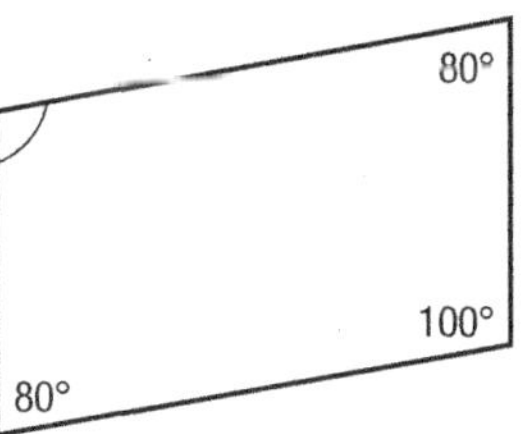

9

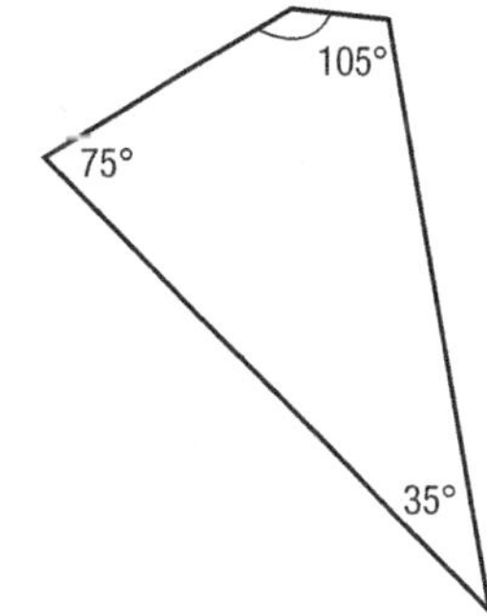

10

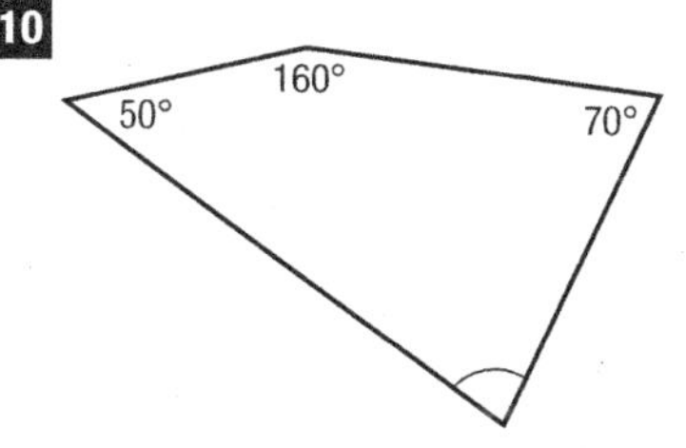

11

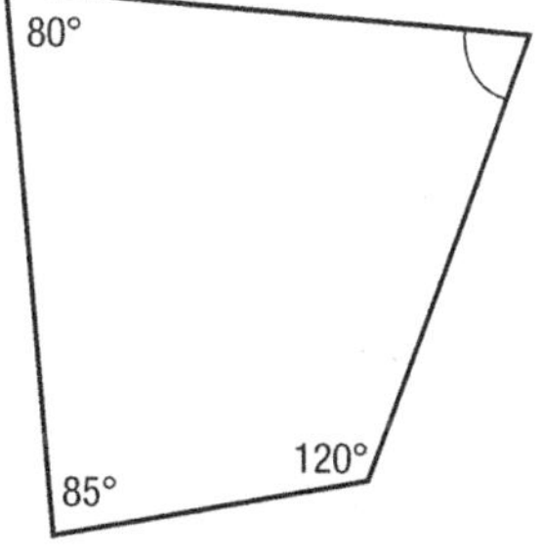

12

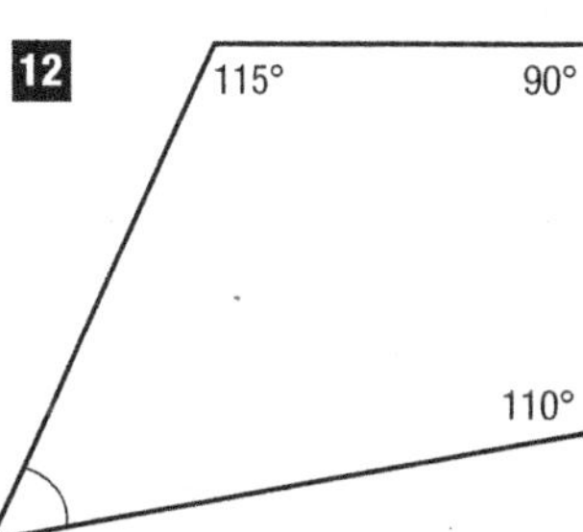

Use straight angles to calculate exterior angles in triangles

An exterior angle is made when the side of a shape is extended to make a straight angle. A straight angle measures 180°.

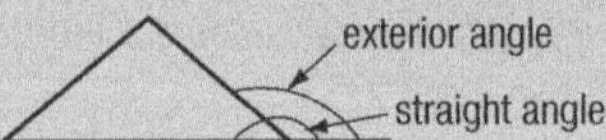

Use what you know about straight angles to calculate the exterior angle that is marked on each triangle.

1 40°

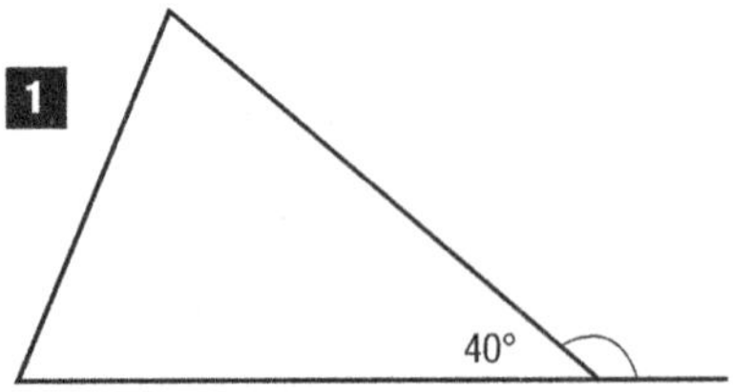

2 45°

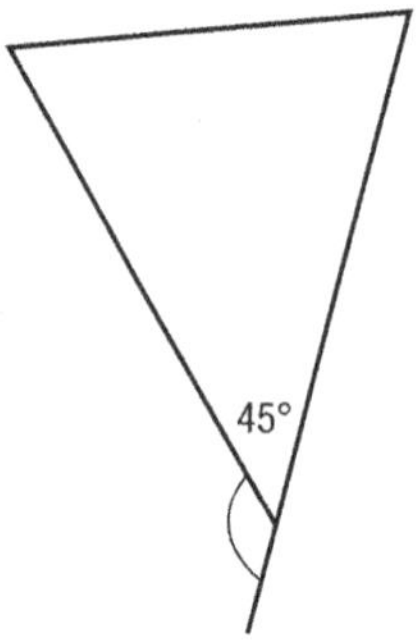

3 55°

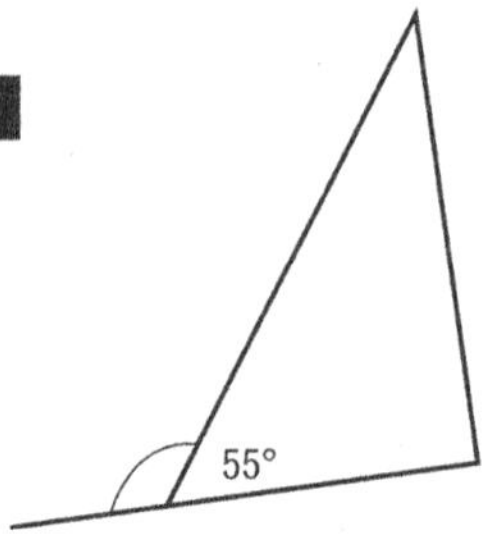

4 105°

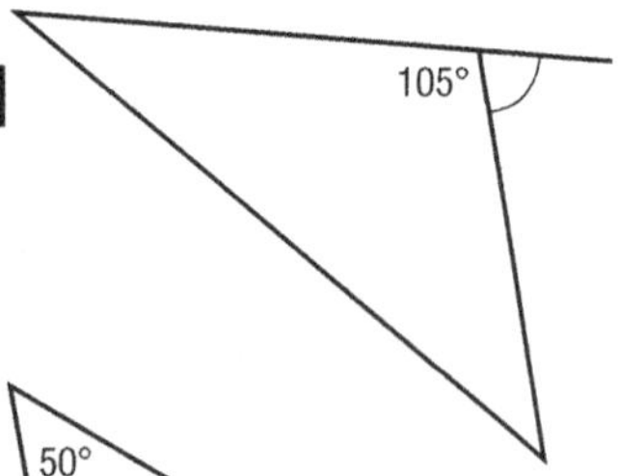

5 125°

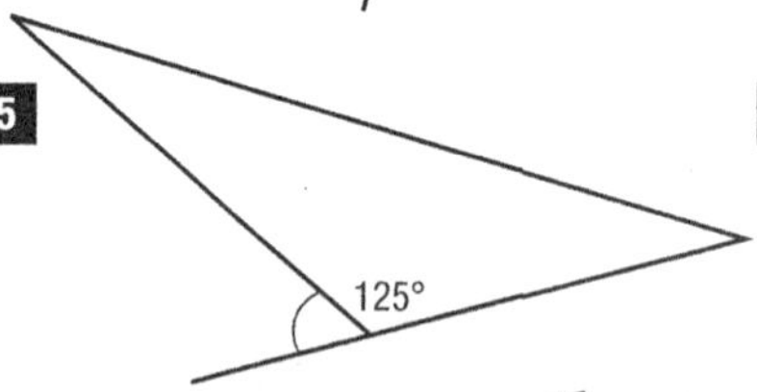

6 60°

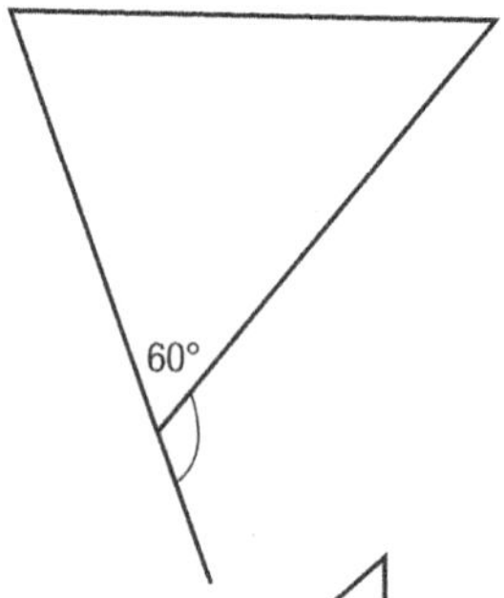

7 50° 70°

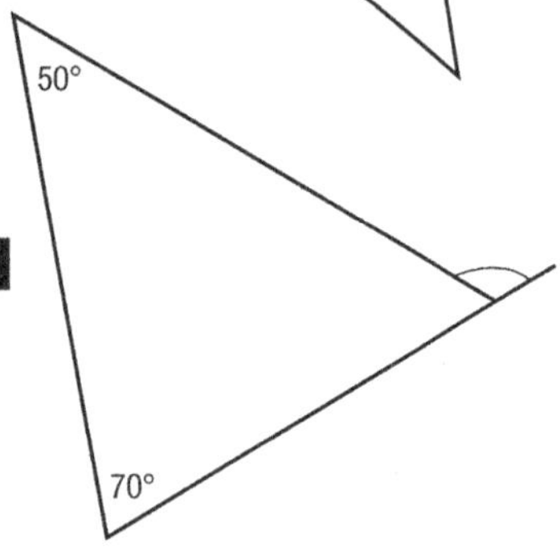

8 60° 45°

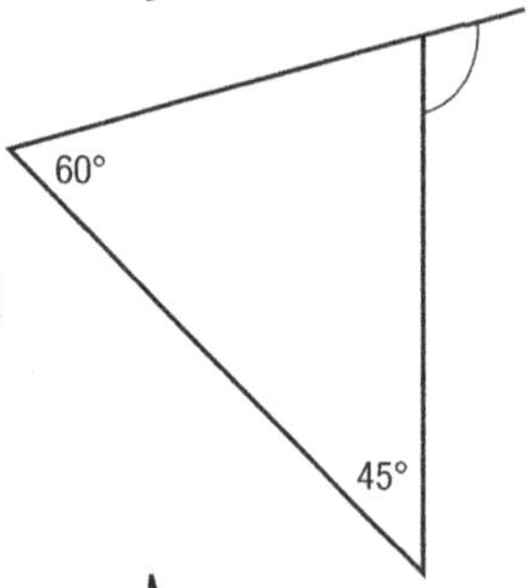

9 40° 100°

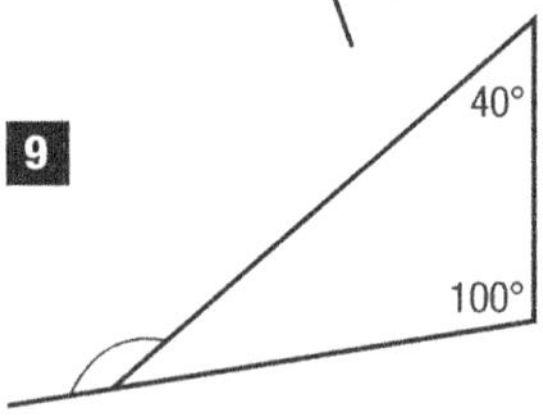

10 85° 40°

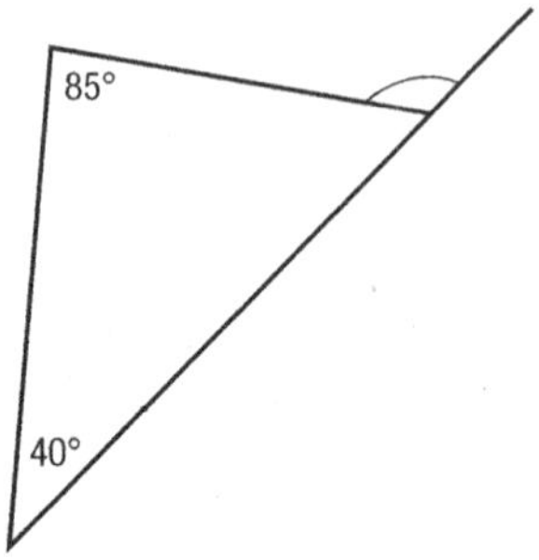

11 25° 75°

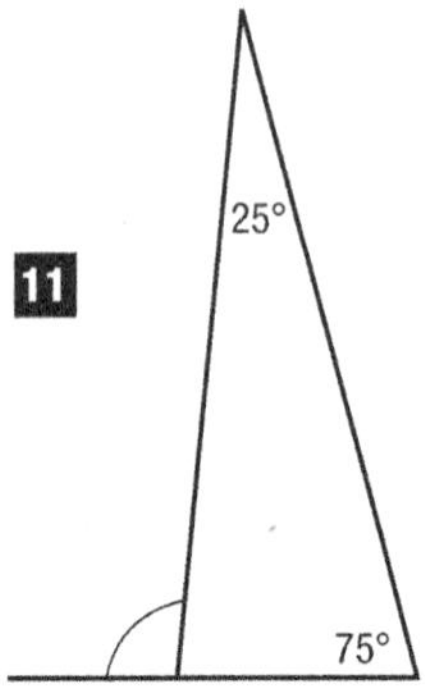

12 65° 60°

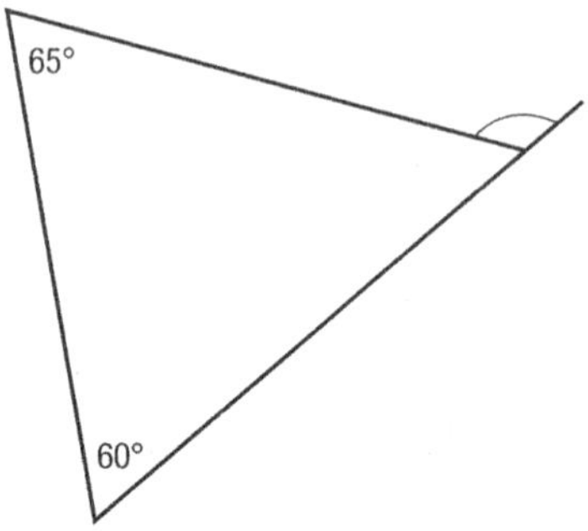

Calculate unknown exterior angles in triangles

Help Box

The exterior angles of a triangle total 360°.

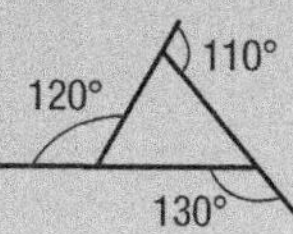

Calculate the missing exterior angle that is marked on each triangle.

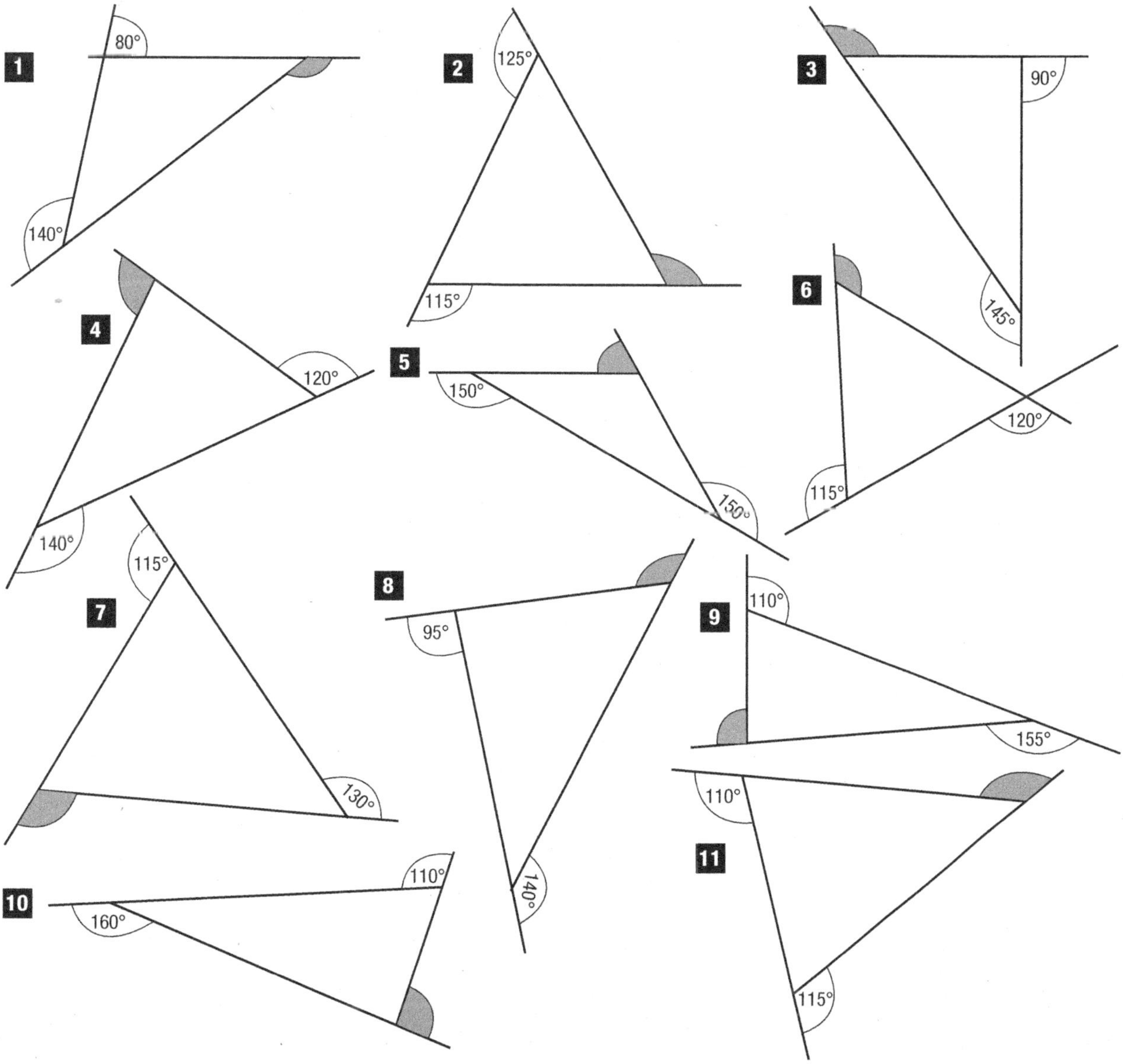

1 80°, 140°

2 125°, 115°

3 90°, 145°

4 120°, 140°

5 150°, 150°

6 120°, 115°

7 115°, 130°

8 95°, 140°

9 110°, 155°

10 110°, 160°

11 110°, 115°

7.2.12 Construct and determine properties of angles

Name and measure angles

1 Write the name (acute, right, obtuse, reflex or straight) of each angle.

2 Estimate the size of each angle.

3 Measure the size of each angle with a protractor.

a

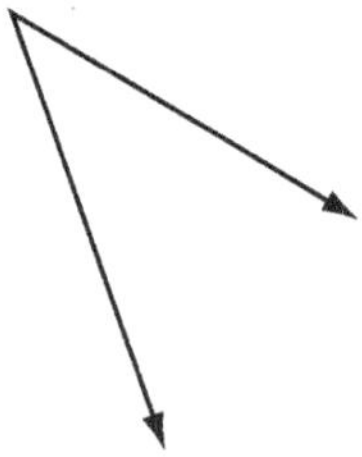

b

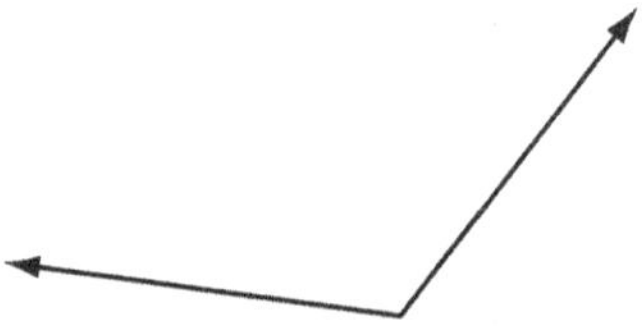

c

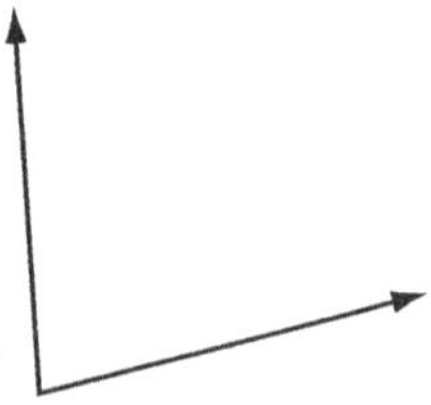

d

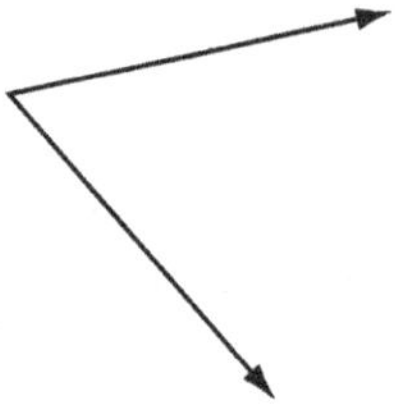

e

f

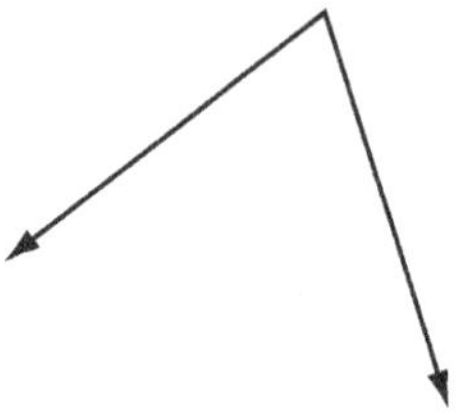

g

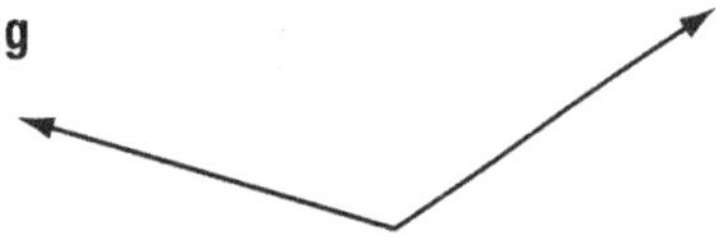

h

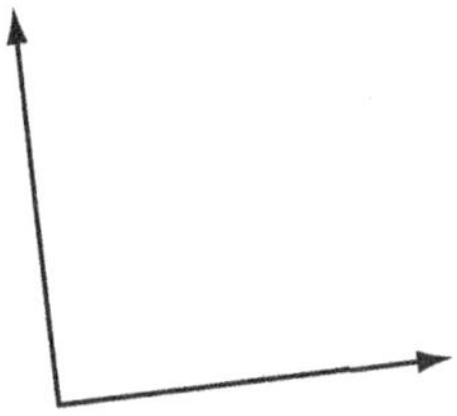

i

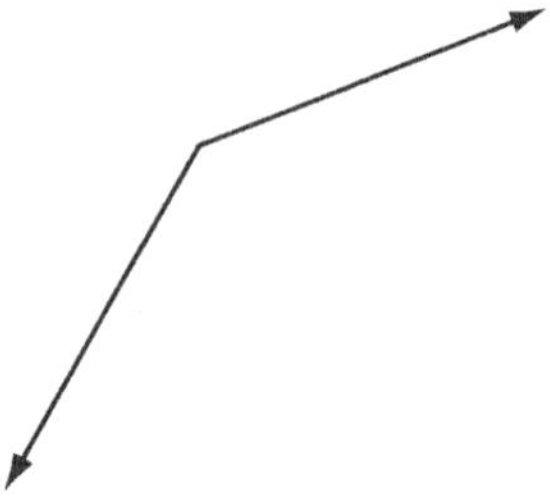

j

k

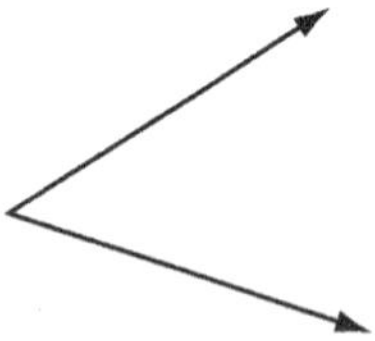

l

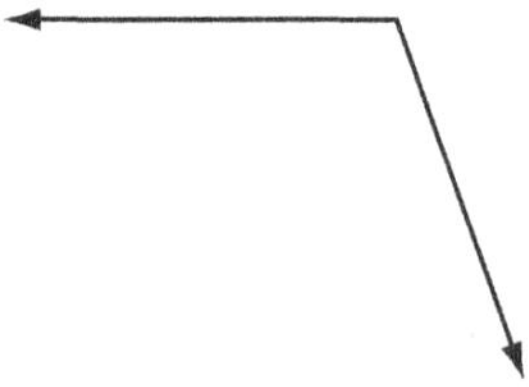

m

n

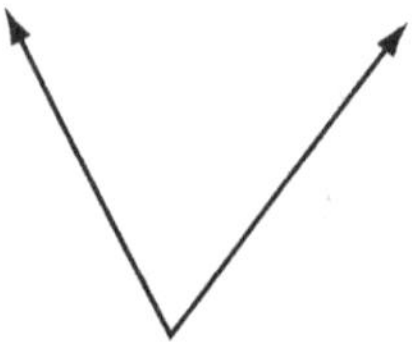

o

Calculate adjacent angles

Remember

Adjacent angles are angles that are next to each other and have a common vertex. In this diagram, ∠WXY is adjacent to ∠YXZ.

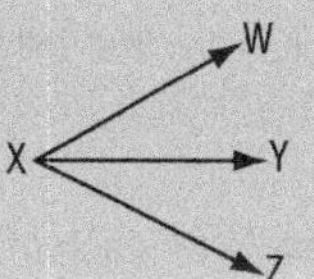

Calculate the size of ∠ABC in each diagram.

1

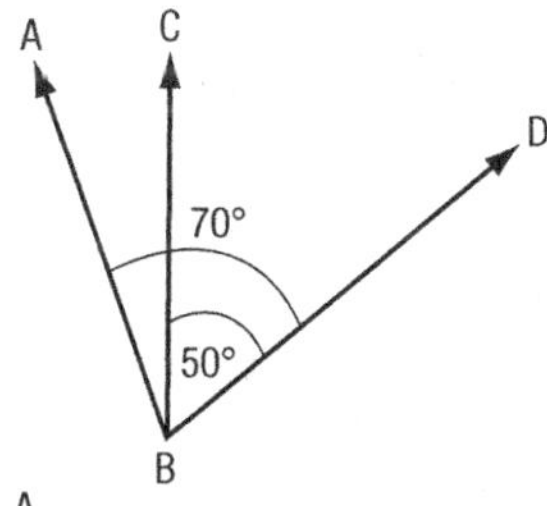

2

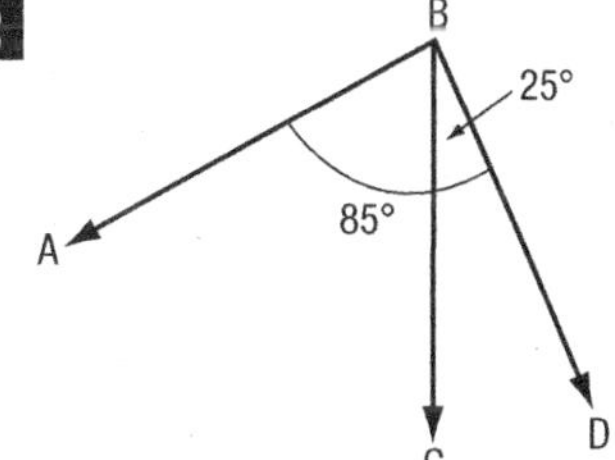

3

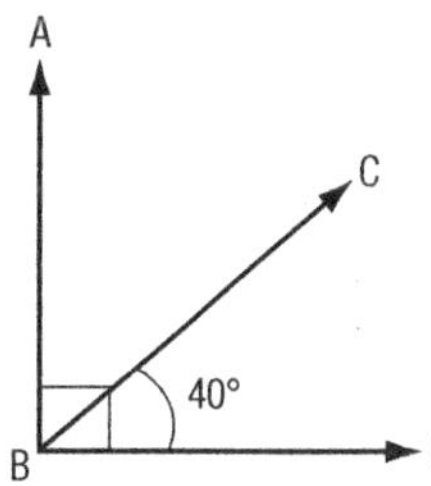

4

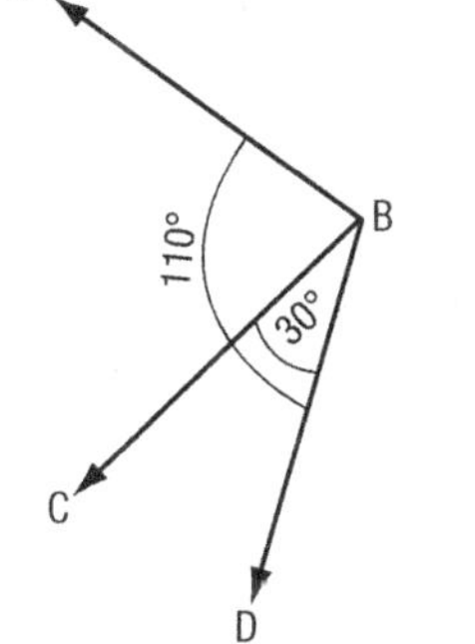

5

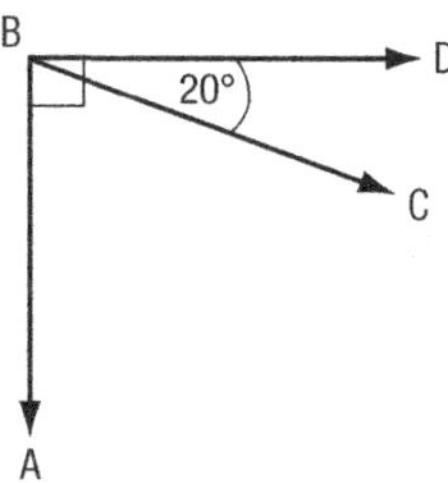

6

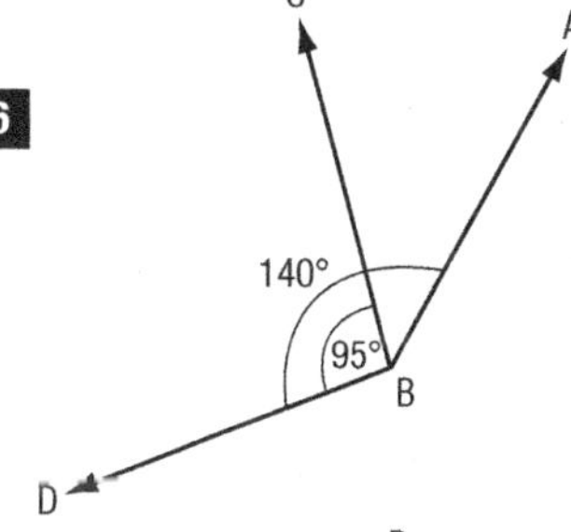

7

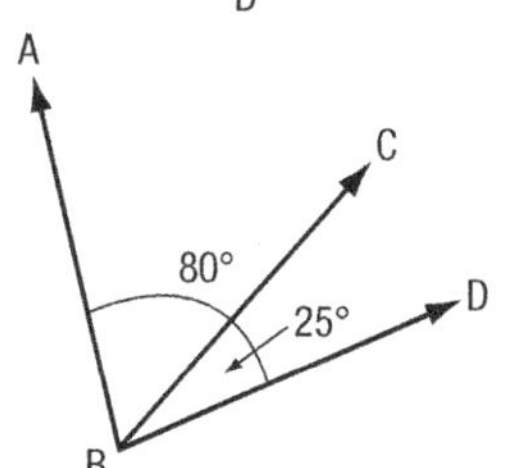

8

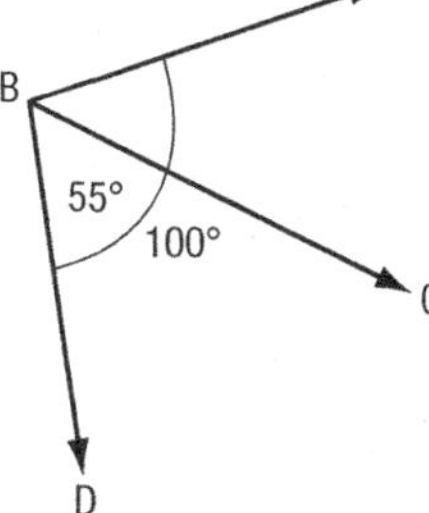

9

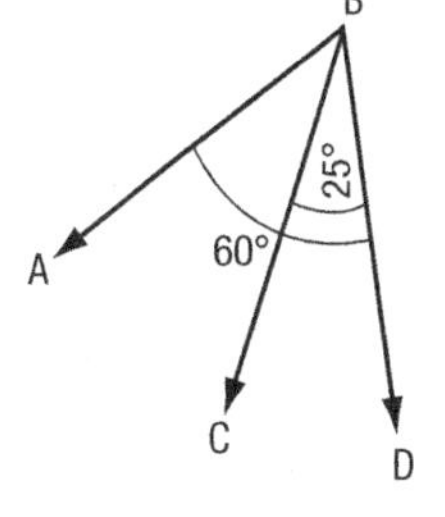

10

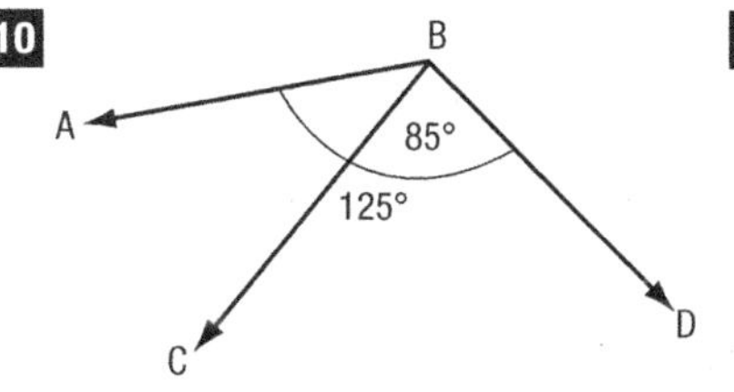

11

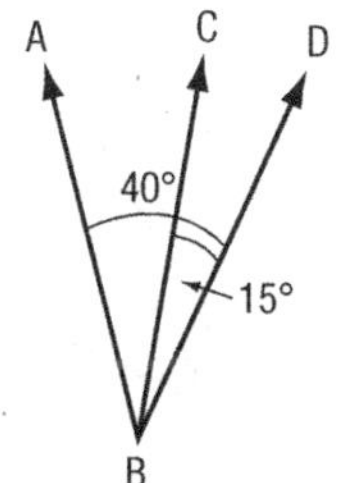

12

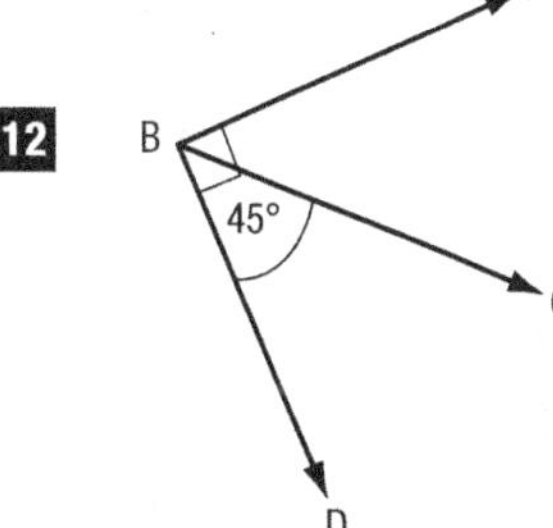

Calculate complementary angles

Remember

Complementary angles are pairs of angles that add together to equal 90°. The angles do not have to be next to each other.

1 What is the complementary angle for each of these angles?

a	55°	**b**	78°	**c**	39°	**d**	21°	**e**	14°	**f**	62°
g	85°	**h**	26°	**i**	7°	**j**	42°	**k**	73°	**l**	59°

2 Calculate the size of ∠JKL in each diagram.

a

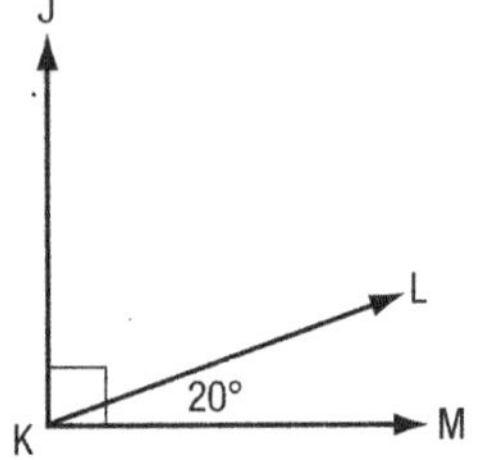

b

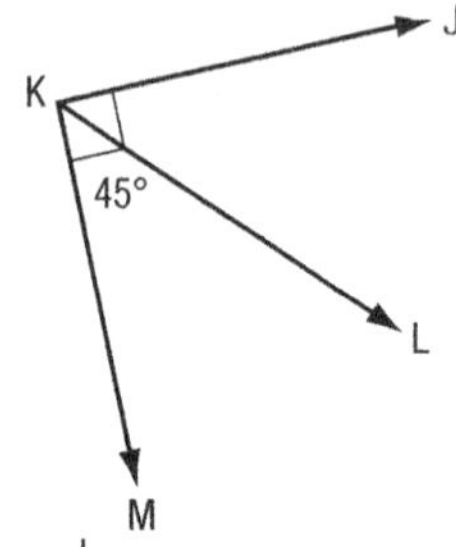

c

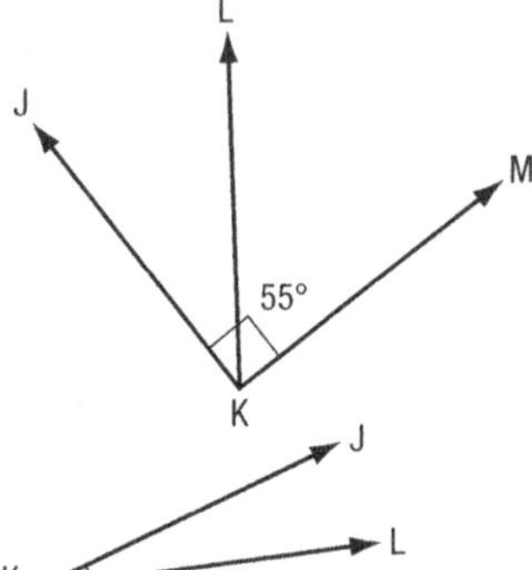

d

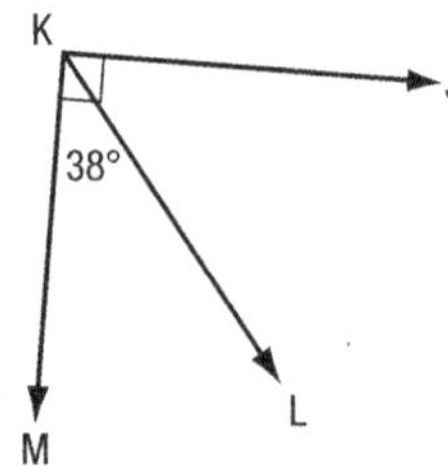

e

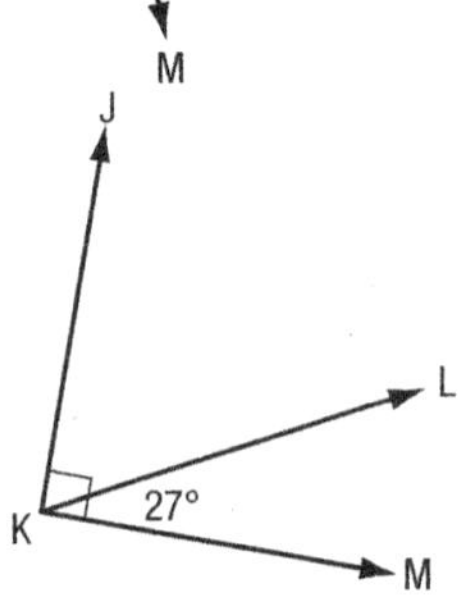

f

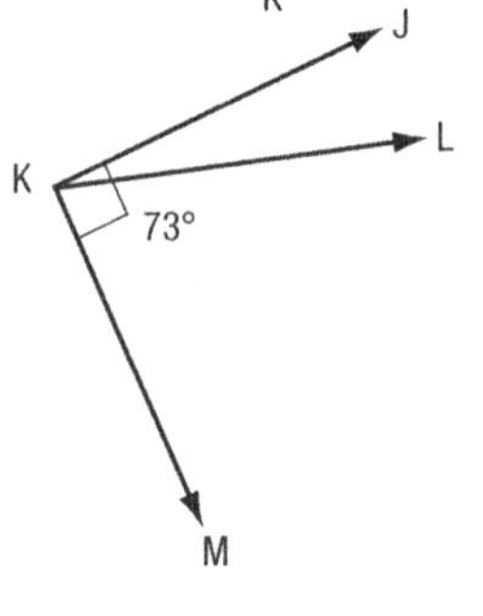

g

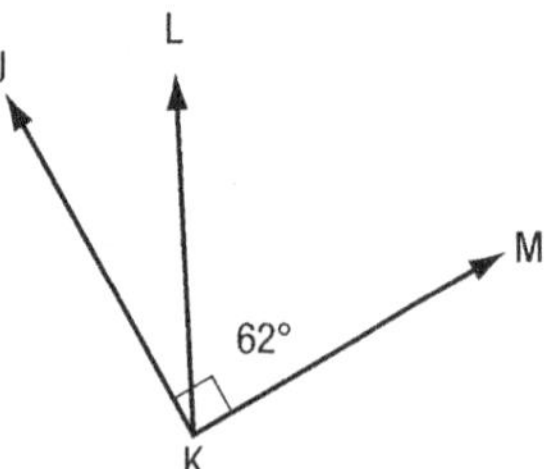

h

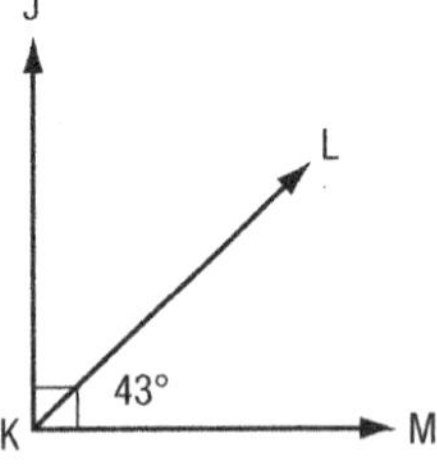

i

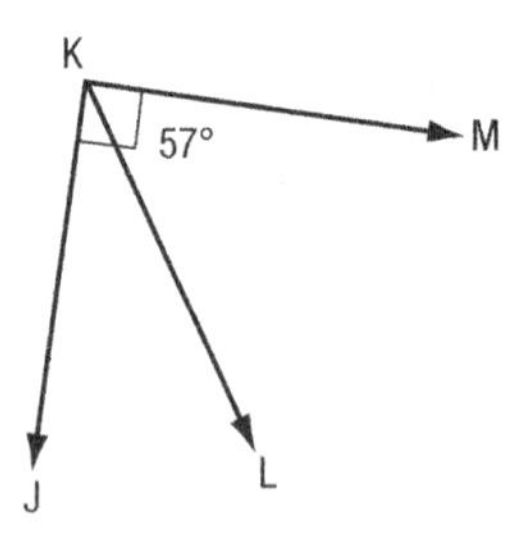

j

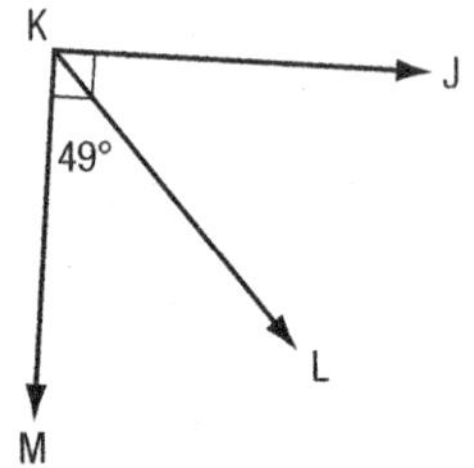

k

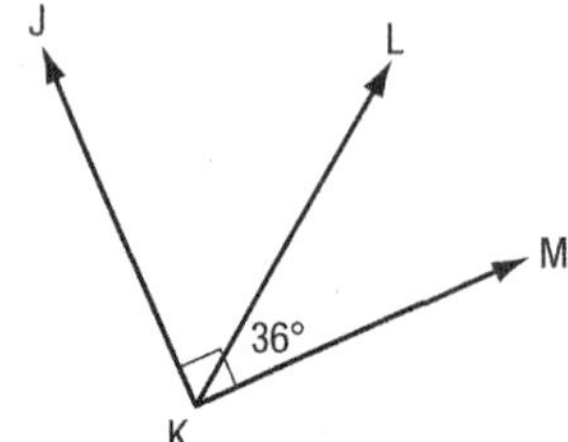

l

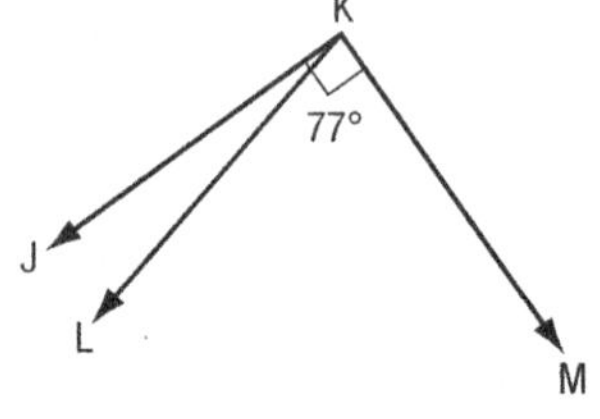

Calculate supplementary angles

Remember

Supplementary angles are pairs of angles that add together to equal 180°. The angles do not have to be next to each other.

1 What is the supplementary angle for each of these angles?

a	90°	**b**	75°	**c**	114°	**d**	135°	**e**	110°	**f**	63°
g	127°	**h**	146°	**i**	87°	**j**	102°	**k**	151°	**l**	59°

2 Calculate the size of ∠DEF in each diagram.

a

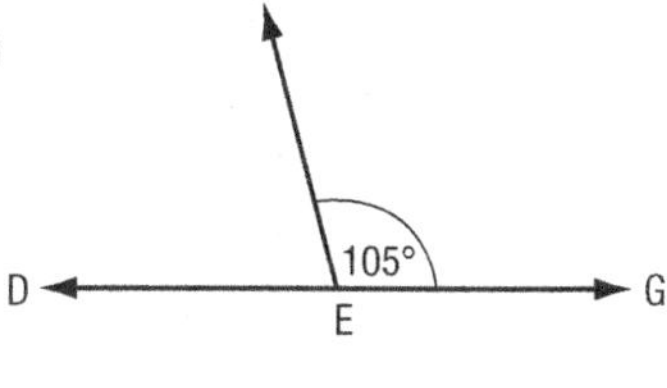

b

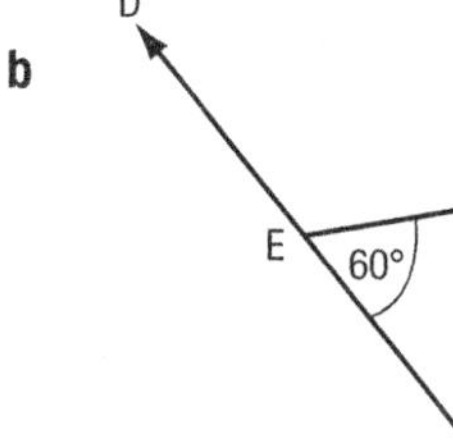

c

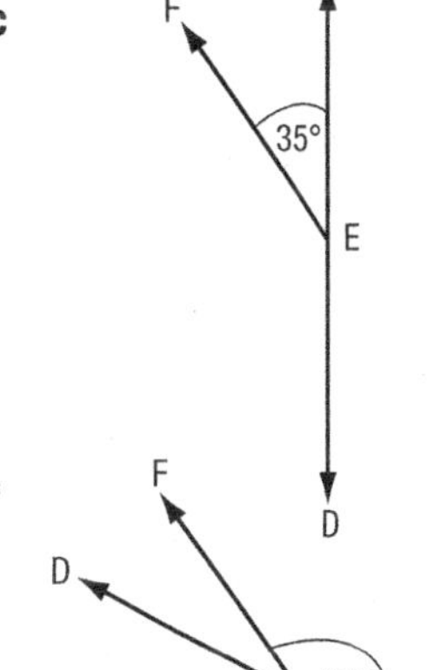

d

D E G
77°
F

e

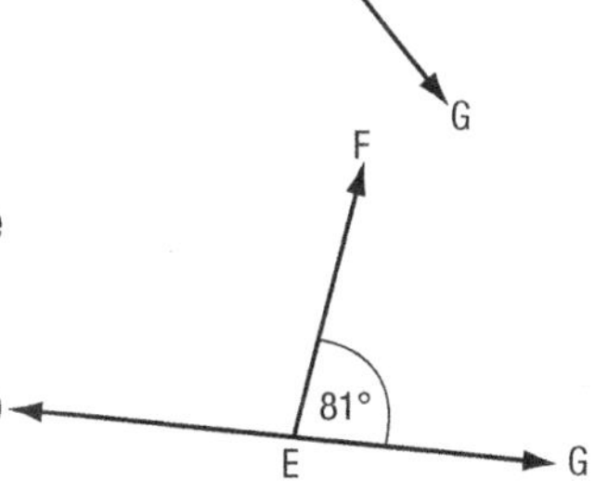

f

F
D
153°
E
G

g

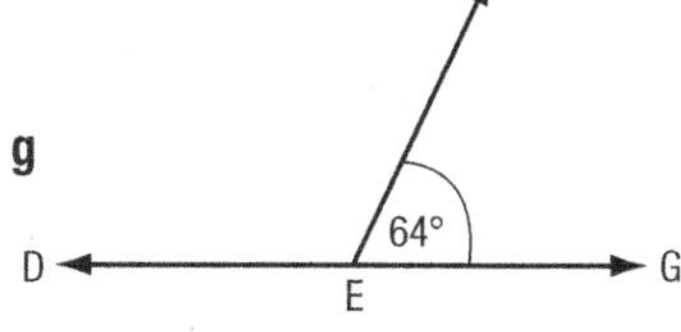

h

D
F
E
112°
G

i

F
41°
G
D
E

j

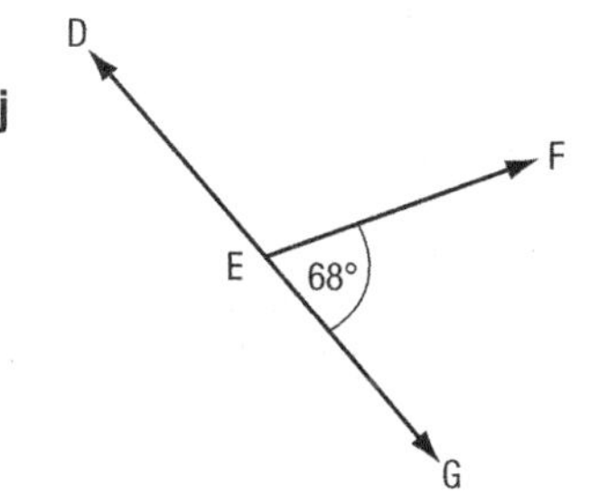

k

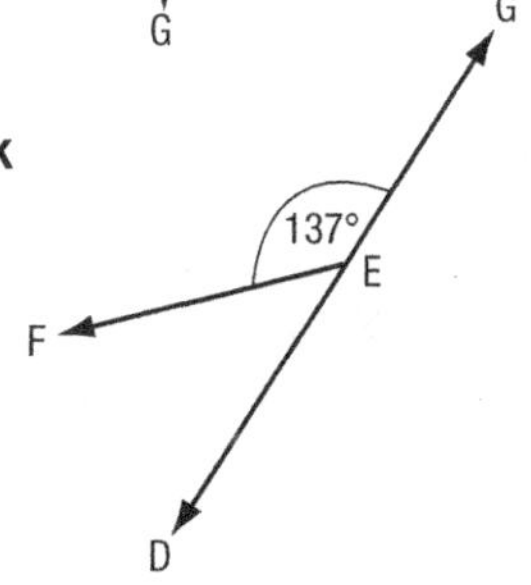

l

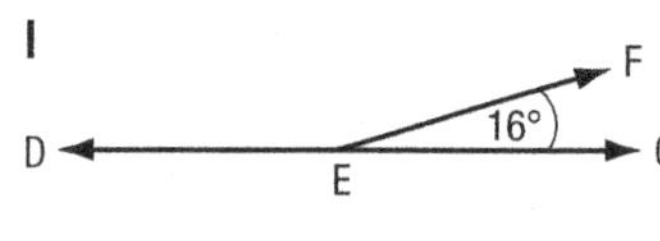

Calculate angles around a point

Remember

Angles around a point make a full turn (or revolution). The total of the angles around a point is 360°.

Calculate the size of the unknown angle in each diagram.

1

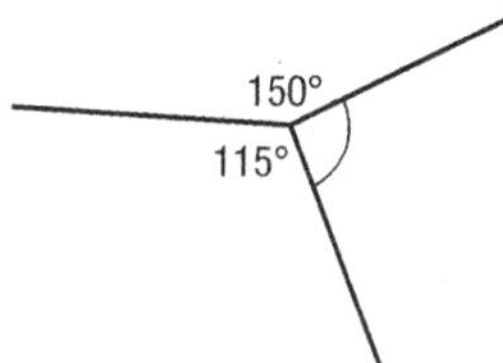

2

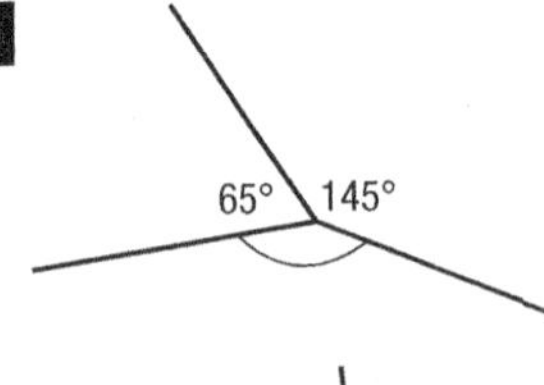

3

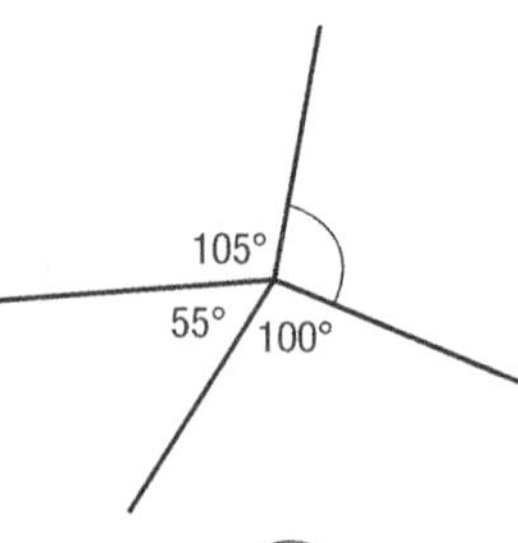

4

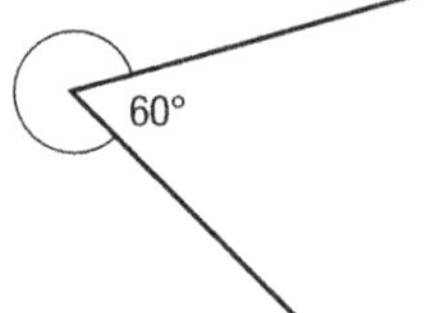

5

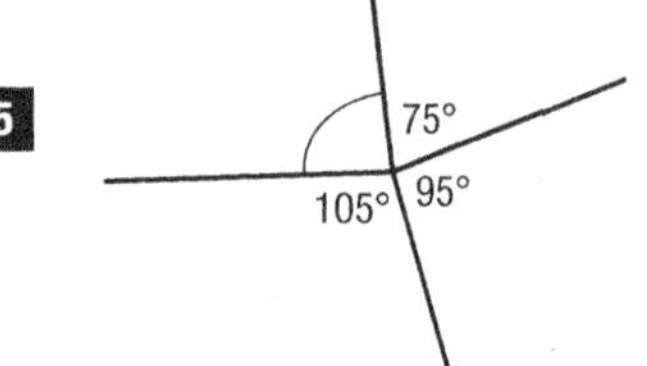

6

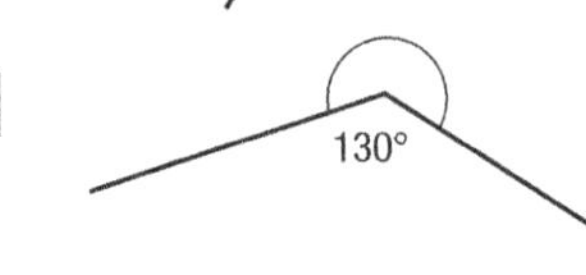

7

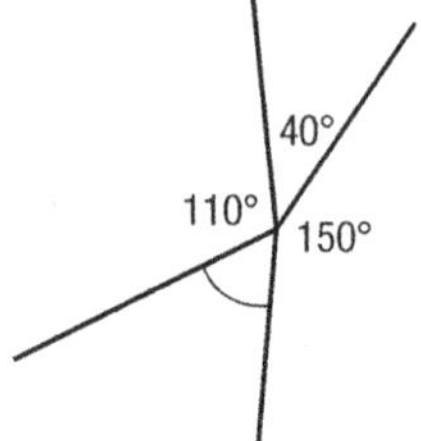

8

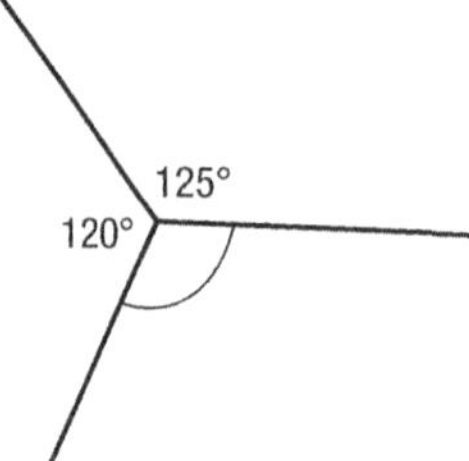

9

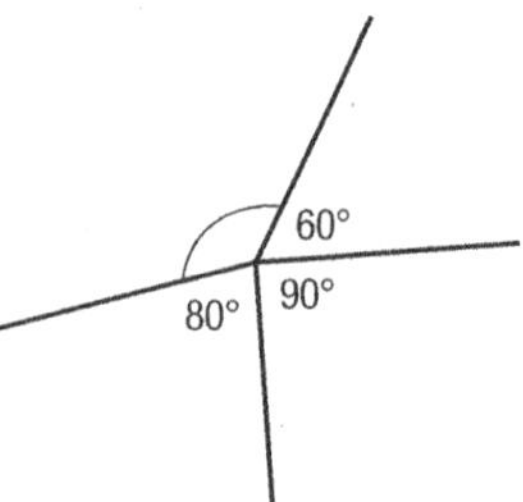

10

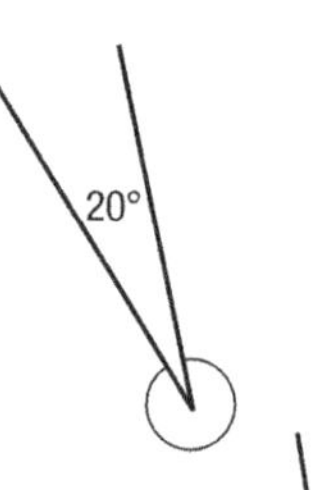

11

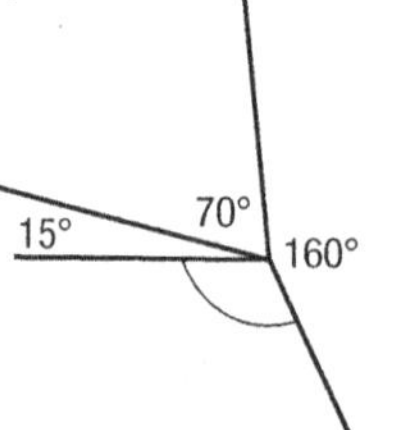

12

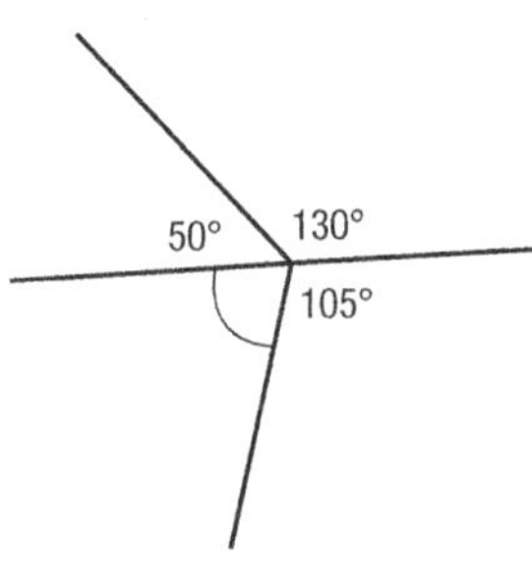

13

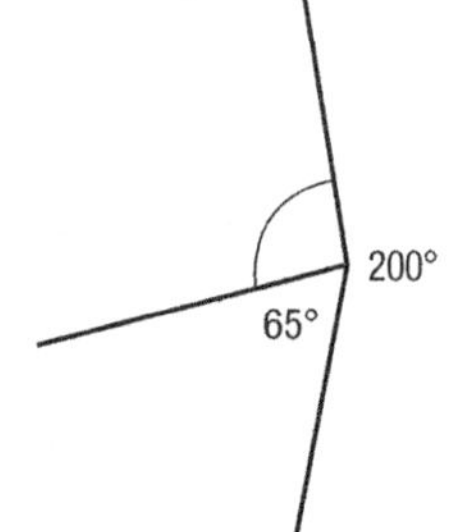

14

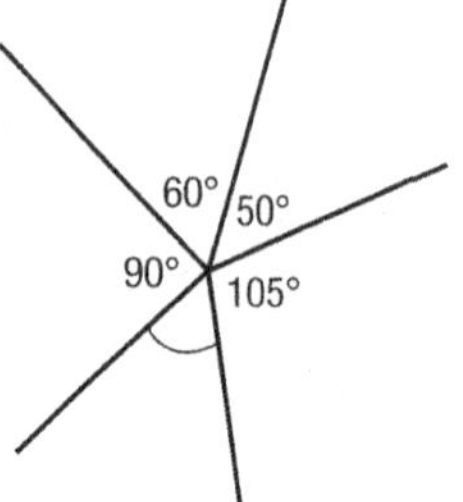

15

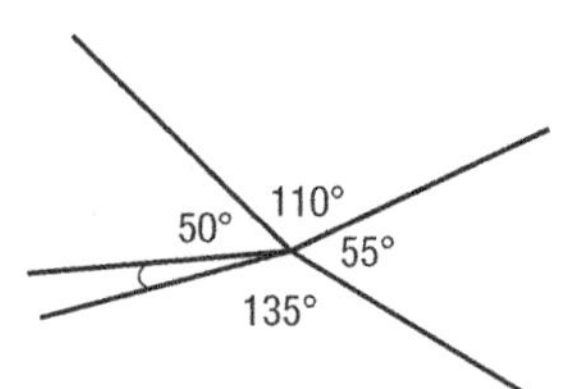

Calculate unknown angles

Use what you have learned about angles to calculate the unknown angles in each diagram.

1

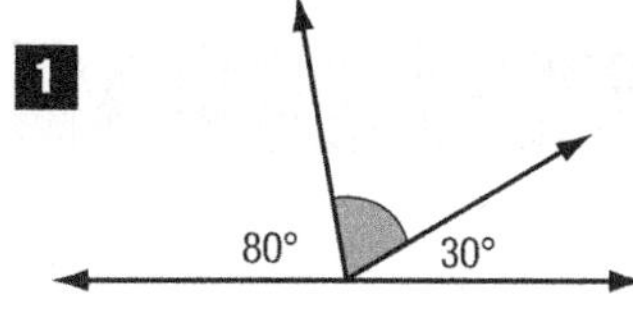

2

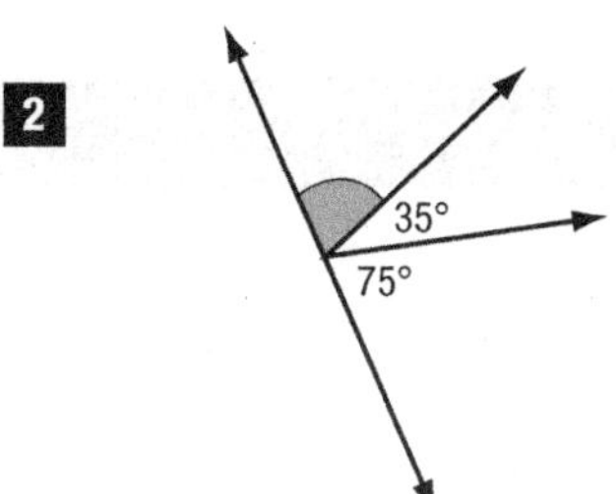

3

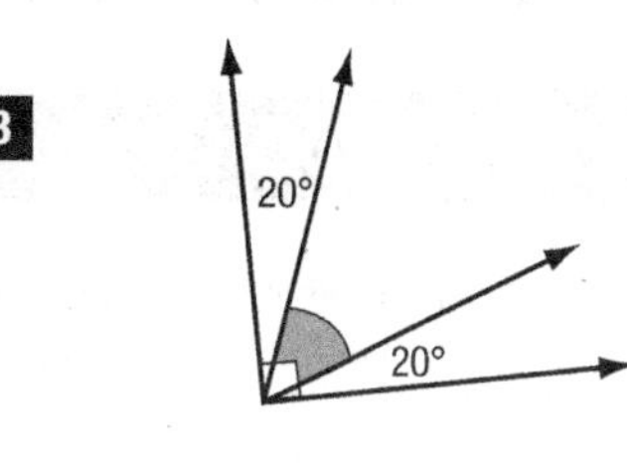

4

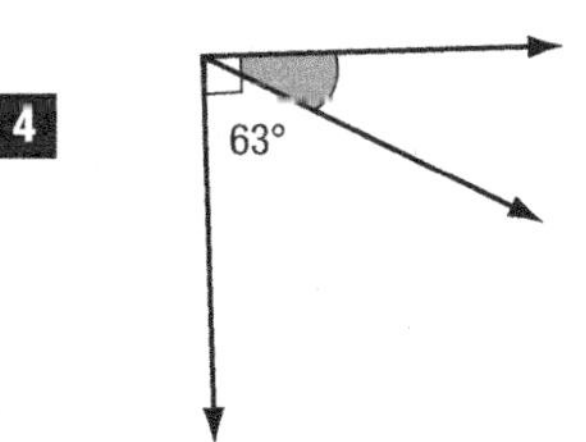

5

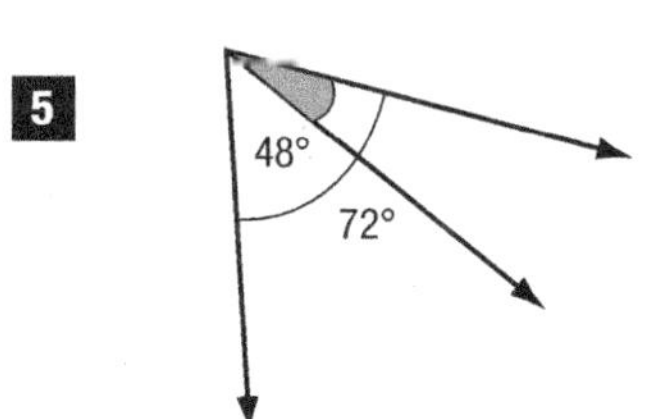

6

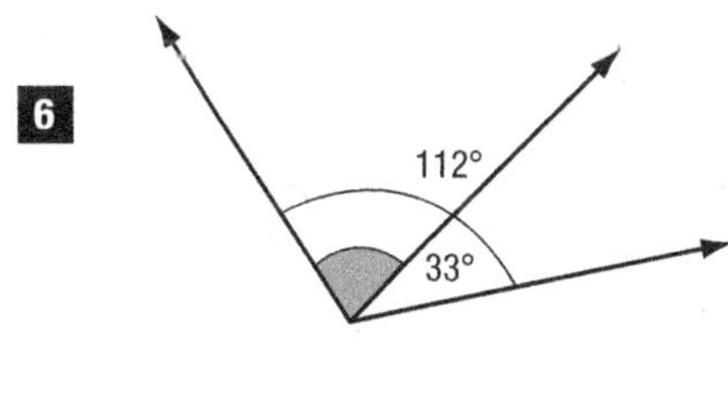

7

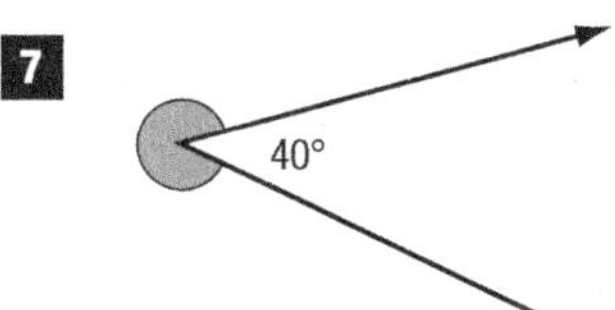

8

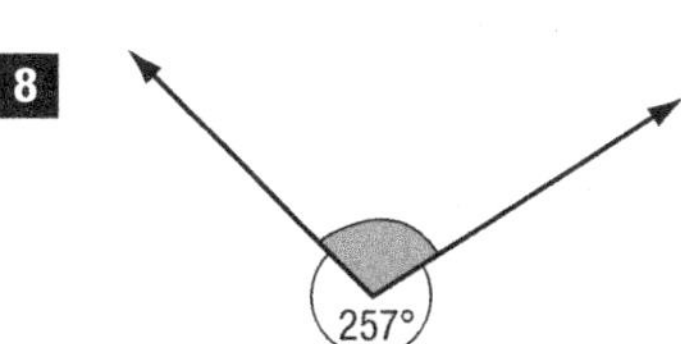

9

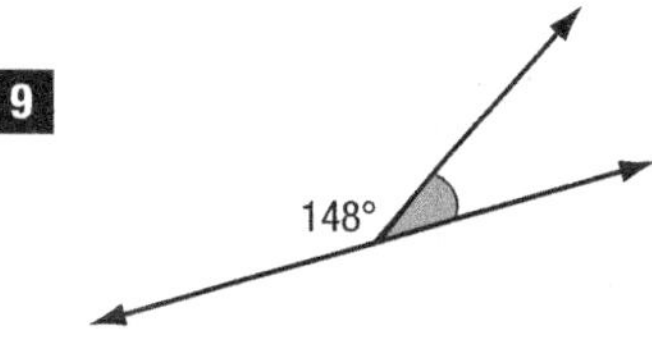

10

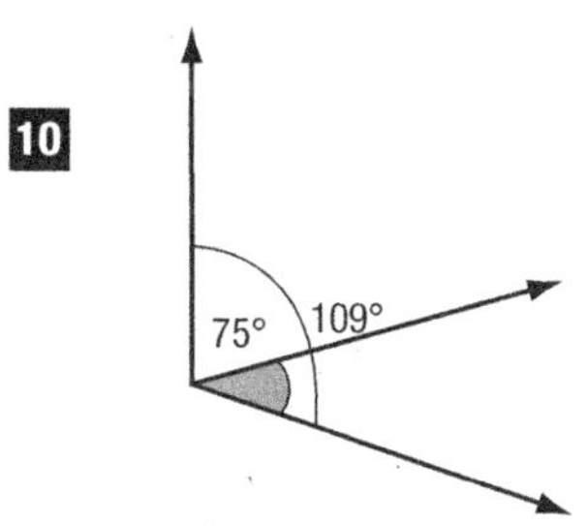

11

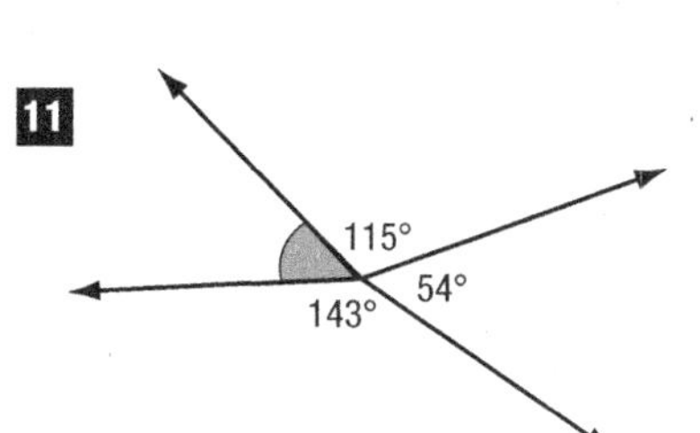

12

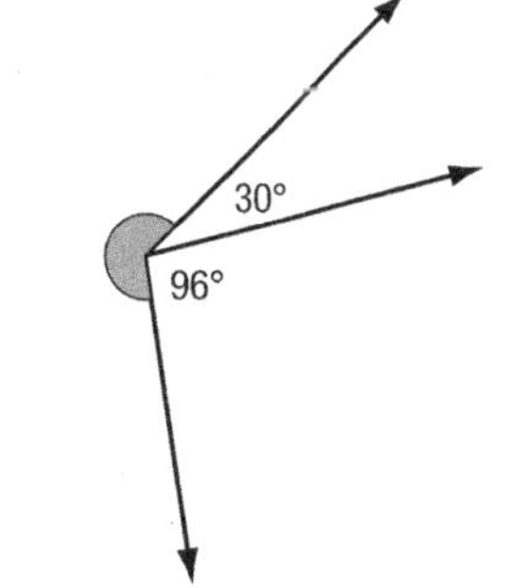

13

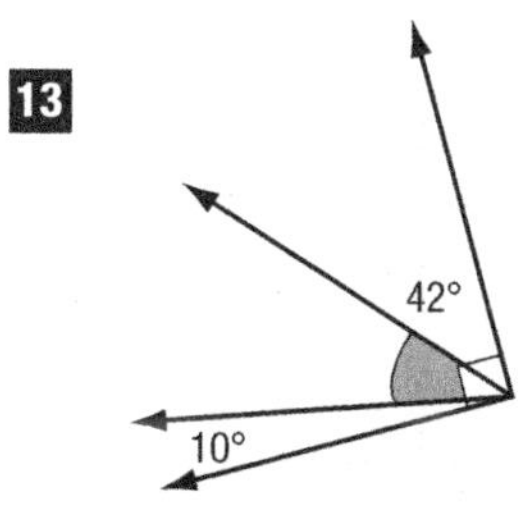

14

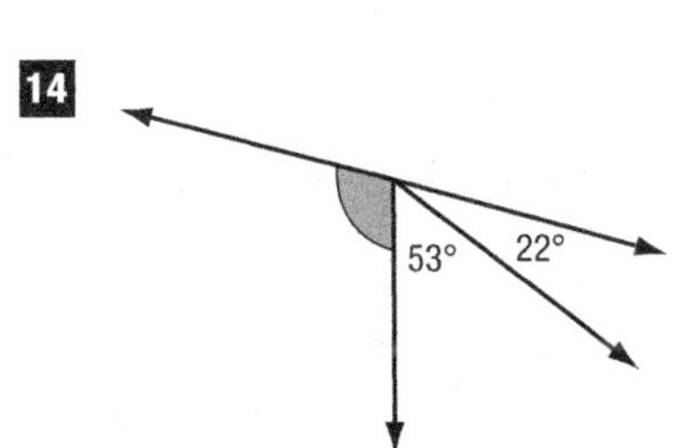

15

130°
70°

7.2.14 Give the direction of a location relative to others as a bearing

State direction as a bearing

Help Box

A compass is used to find a direction or bearing.
The bearing of a point is the angle measured in degrees in a clockwise direction from the north line to the line joining the centre of the compass. For example, the compass drawn here shows the bearing of point A is 050° and the bearing of point B is 150° (90° + 60°).

All bearings are given with three digits: 050° not 50°.

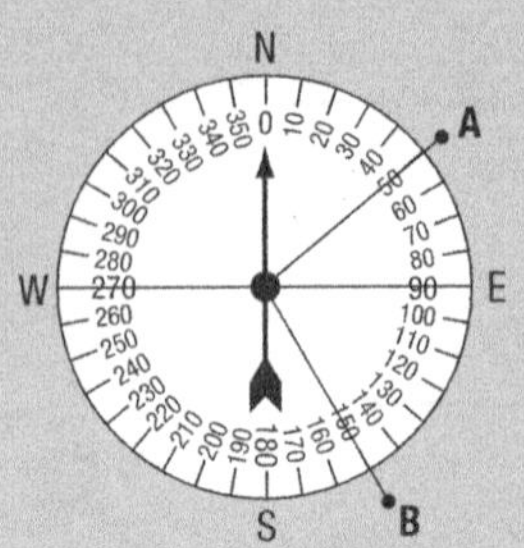

State the bearing of point K in each diagram.

1

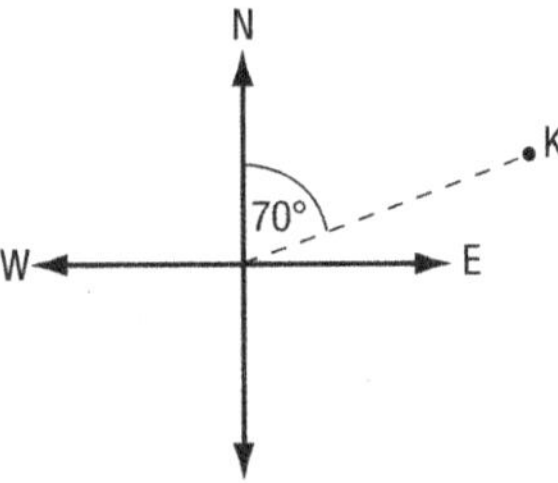

2

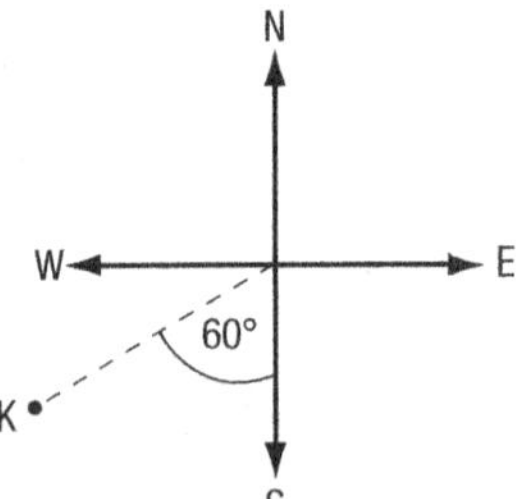

3

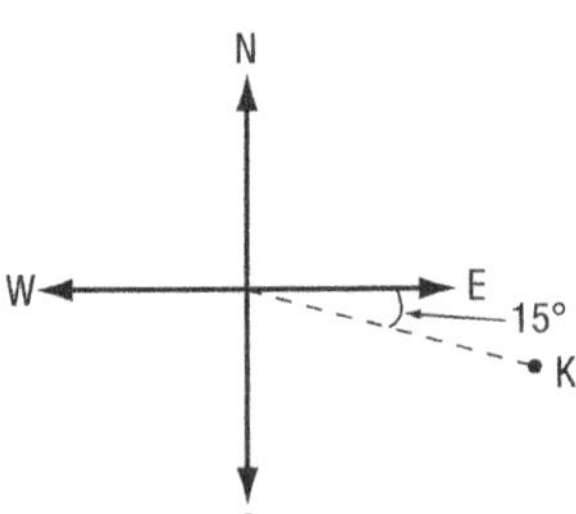

4

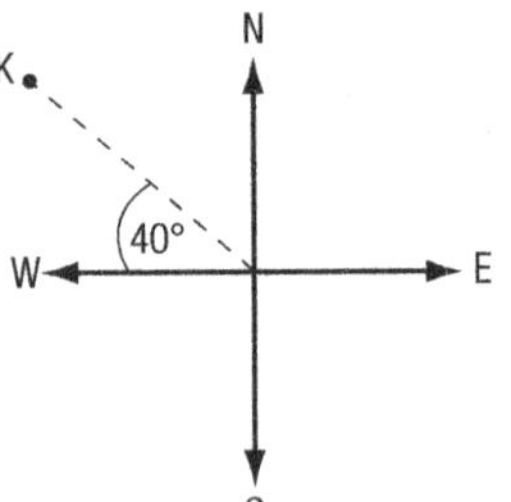

5

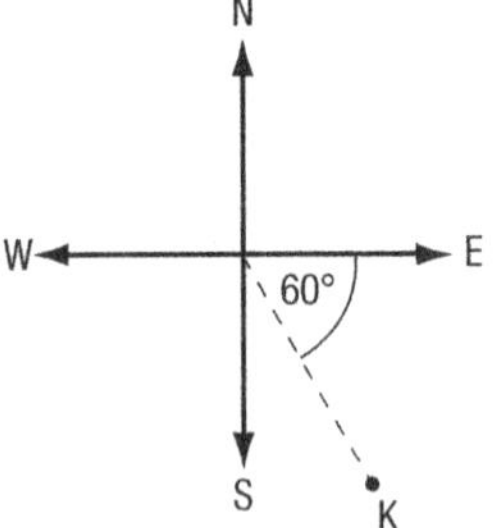

6

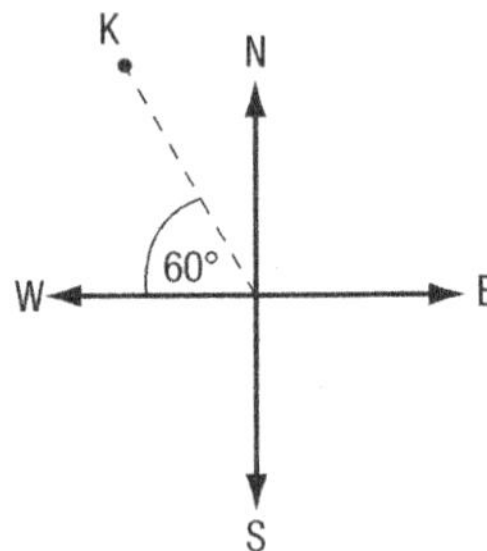

7

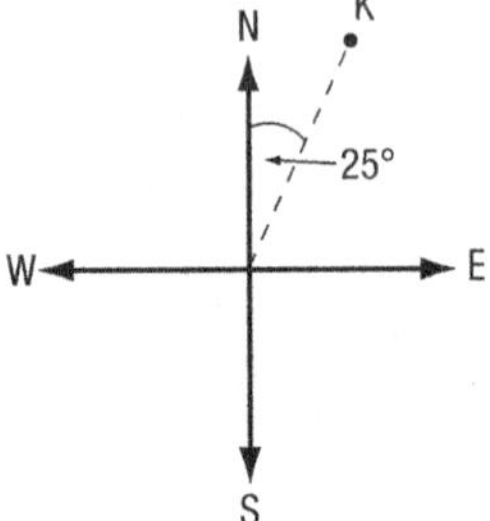

8

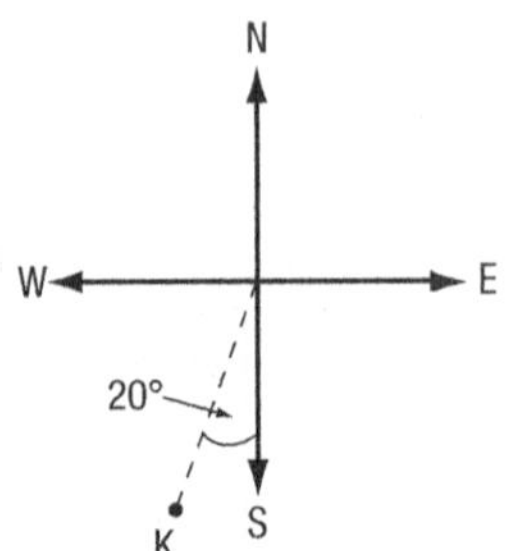

9

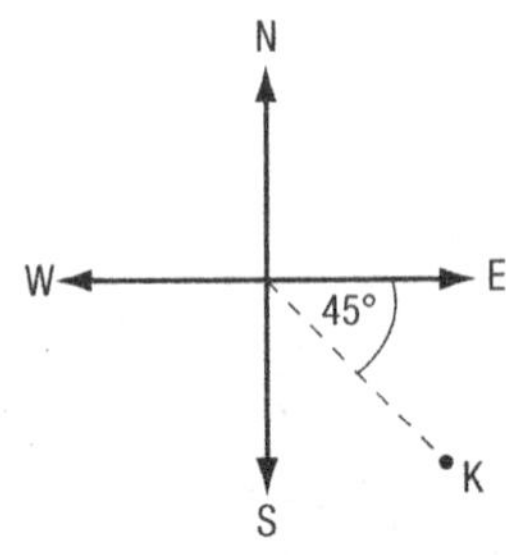

State direction with compass points

Help Box

Direction can be given in degrees. It is usually stated as the number of degrees east or west of the north-south line. For example, the diagram drawn here shows A is N45°E, which means the direction is 45° east of north; B is S70°E, which means the direction is 70° east of south; C is S50°W, which means the direction is 50° west of south; D is N80°W, which means the direction is 80° west of north.

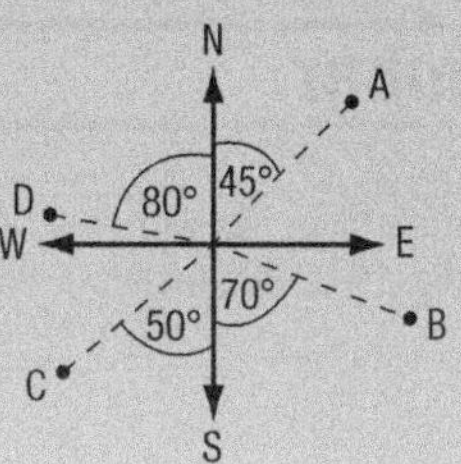

1 In each diagram write the direction of A, B, C and D from the north-south line.

a

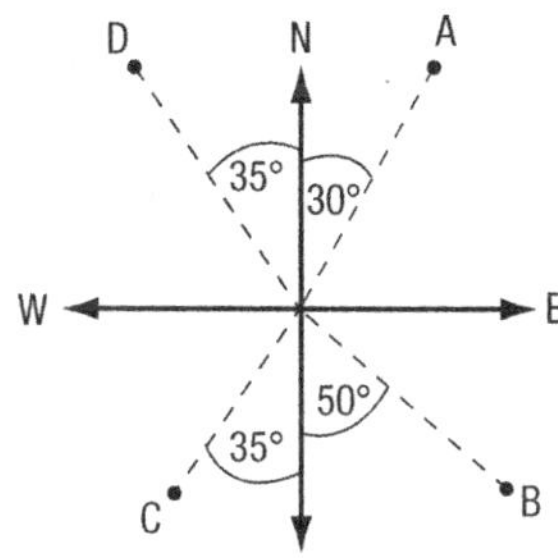

b

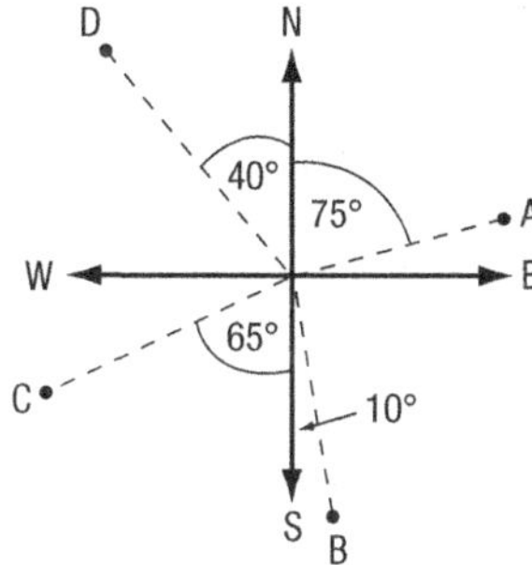

c

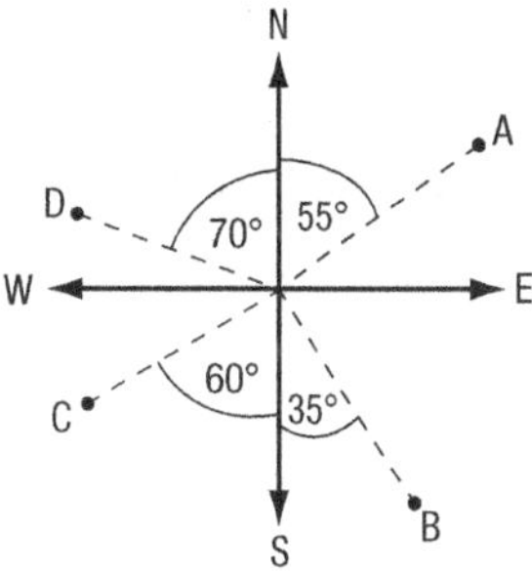

d

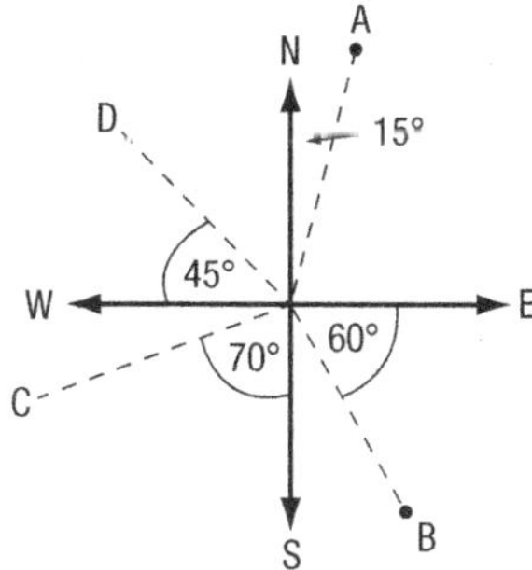

e

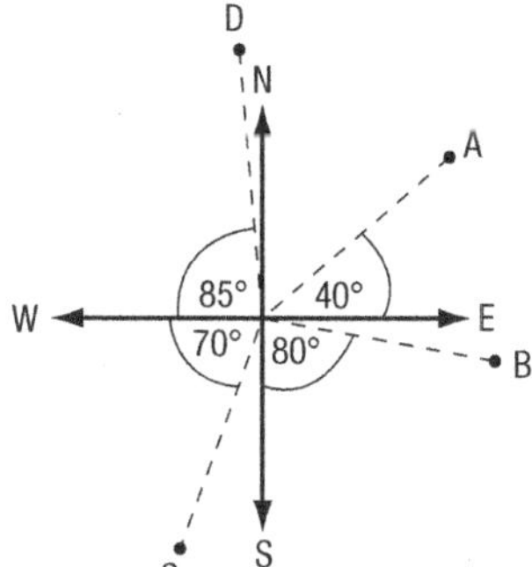

f

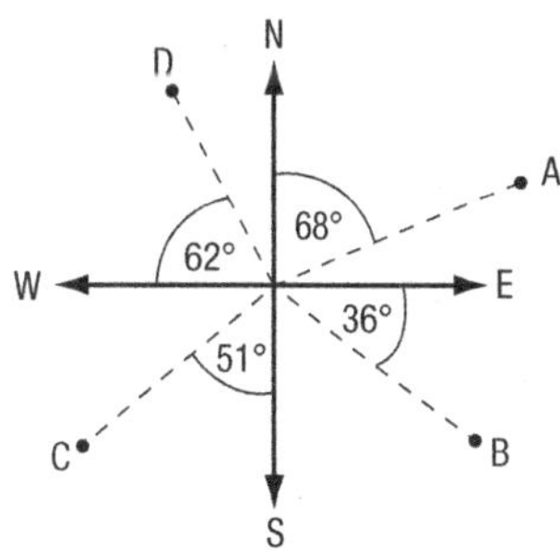

2 Write each bearing as a direction. For example, a bearing of 160° would be S20°E.

a	125°	**b**	035°	**c**	150°	**d**	245°	**e**	080°	**f**	215°
g	330°	**h**	190°	**i**	275°	**j**	056°	**k**	172°	**l**	310°
m	104°	**n**	298°	**o**	339°	**p**	027°				

7.2.15 Make maps having scales and keys from suitable data

Understand scale as a ratio

Remember

The scale on maps is sometimes written as a ratio. It tells us the relationship between the distances on the map and the distances in real life. For example, 1:100 000 means 1 cm on the map represents 1 km in real life, since 1 km = 100 000 cm.

1 Convert these measurements to the same unit of measurement and rewrite the scale as a ratio.

a	1 cm = 5 km	**b**	1 cm = 0.5 km	**c**	1 cm = 3 km	**d**	1 cm = 10 m
e	1 cm = 6 m	**f**	1 cm = 2.5 m	**g**	2 cm = 1 km	**h**	2 cm = 500 m
i	2 cm = 5 m	**j**	5 cm = 10 km	**k**	5 cm = 2 km	**l**	5 cm = 100 m

2 The scale used on a map is 1:5000. Calculate the real distances in metres for these distances measured from the map.

a	2 cm	**b**	4 cm	**c**	7 cm	**d**	10 cm	**e**	5 mm	**f**	3.5 cm
g	8.5 cm	**h**	1.5 cm	**i**	12 cm	**j**	15.5 cm	**k**	6.2 cm	**l**	11.6 m

3 The scale used on a map is 1:200 000. Calculate the real distances in kilometres for these distances measured from the map.

a	3 cm	**b**	8 cm	**c**	10 cm	**d**	6 cm	**e**	0.5 cm	**f**	25 cm
g	16 cm	**h**	30 cm	**i**	7.5 cm	**j**	12.5 cm	**k**	11.25 cm	**l**	24.75 cm

4 The scale used on a map is 1:2 500 000. Calculate the real distances in kilometres for these distances measured from the map.

a	4 cm	**b**	7 cm	**c**	12 cm	**d**	5 cm	**e**	1.5 cm	**f**	15 cm
g	3.5 cm	**h**	10.5 cm	**i**	6.2 cm	**j**	2.8 cm	**k**	11.4 cm	**l**	9.6 cm

5 If the scale used on a map is 1:10 000, how many centimetres would represent these actual distances?

a	200 m	**b**	500 m	**c**	50 m	**d**	150 m	**e**	1 km	**f**	5 km
g	850 m	**h**	1250 m	**i**	2.5 km	**j**	0.8 km	**k**	230 m	**l**	610 m

6 If the scale used on a map is 1:150 000, how many centimetres would represent these actual distances?

a	1.5 km	**b**	6 km	**c**	12 km	**d**	10.5 km	**e**	15 km	**f**	750 m
g	30 km	**h**	22.5 km	**i**	37.5 km	**j**	25.5 km	**k**	19.5 km	**l**	33 km

7 If the scale used on a map is 1:6000, how many centimetres would represent these actual distances?

a	120 m	**b**	30 m	**c**	150 m	**d**	240 m	**e**	0.6 km	**f**	1.8 km
g	480 m	**h**	630 m	**i**	750 m	**j**	1.5 km	**k**	390 m	**l**	660 m

7.2.16 Use a four-quadrant number plane

Read and record ordered pairs on a number plane

Remember

The first number in a pair gives the number of units left or right of the origin along the x-axis. The second number in a pair gives the number of units above or below the origin along the y-axis.

1 Give the ordered pairs for the following letters on the number plane.

a	P	**b**	S	**c**	M
d	B	**e**	K	**f**	R
g	E	**h**	T	**i**	H
j	N	**k**	X	**l**	Q

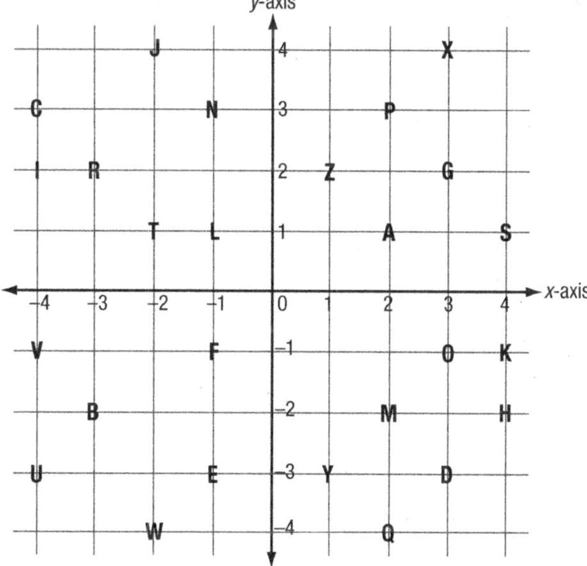

2 What letters are at the following points?

a	(2, 1)	**b**	(3, –3)	**c**	(–1, 1)
d	(–1, –1)	**e**	(–2, –4)	**f**	(3, 2)
g	(3, –1)	**h**	(–4, 3)	**i**	(–2, 4)
j	(–4, –3)	**k**	(1, 2)	**l**	(1, –3)

3 What Papua New Guinea place names do the following sets of ordered pairs spell out?

a (–3, 2) (2, 1) (–3, –2) (2, 1) (–4, –3) (–1, 1)

b (4, –1) (–4, 2) (2, –2) (–3, –2) (–1, –3)

c (2, –2) (2, 1) (3, –3) (2, 1) (–1, 3) (3, 2)

d (3, –3) (2, 1) (–3, 2) (–4, –3)

e (–2, –4) (–1, –3) (–2, –4) (2, 1) (4, –1)

f (2, –2) (–1, –3) (–1, 3) (3, –3) (–4, 2)

g (4, –1) (–1, –3) (–3, 2) (–1, –3) (2, –2) (2, 1)

h (2, 3) (3, –1) (2, –2) (–4, 2) (3, –1)

i (2, 1) (–1, 1) (3, –1) (–2, 1) (2, 1) (–4, –3)

j (–2, 1) (–4, –3) (–1, –1) (–4, 2)

k (–2, 1) (–1, –3) (–1, 1) (–1, –3) (–1, –1) (3, –1) (2, –2) (–4, 2) (–1, 3)

l (2, 3) (3, –1) (2, 3) (3, –1) (–1, 3) (3, –3) (–1, –3) (–2, 1) (–2, 1) (2, 1)

m (4, –1) (–4, –3) (–1, 3) (3, –3) (–4, 2) (2, 1) (–2, –4) (2, 1)

n (–1, 3) (2, 1) (2, –2) (2, 1) (–2, 1) (2, 1) (–1, 3) (2, 1) (–4, 2)

o (2, 1) (2, –2) (–3, –2) (–4, –3) (–1, 3) (–2, 1) (–4, 2)

4 Write the set of ordered pairs that spell these country names.

a	AUSTRALIA	**b**	NEW ZEALAND	**c**	INDONESIA	**d**	THAILAND	**e**	PHILIPPINES
f	JAPAN	**g**	CHINA	**h**	VIETNAM	**i**	PAKISTAN	**j**	MALAYSIA
k	MOROCCO	**l**	ETHIOPIA	**m**	ARGENTINA	**n**	VENEZUELA	**o**	BRAZIL

Assessment — Space and Shape

1 Convert the lengths in each set to the same unit and then order them from shortest to longest.

a 8.4 m 835 cm 8405 mm   b 80 500 cm 80 m 0.8 km

c 2.03 km 2300 m 20 300 cm   d 475 mm 0.47 m 4.7 cm

2 Find the circumference of circles that have the following diameters.

a 12 cm b 31 cm c 27 cm d 19 cm e 23.5 cm f 16.8 cm

3 Find the circumference of circles that have the following radii.

a 25 cm b 18 cm c 11.25 cm d 20.5 cm e 9.75 cm f 12.3 cm

4 Find the diameter of circles that have the following circumferences.

a 25.12 cm b 17.27 cm c 36.11 cm d 7.85 cm

5 Use the given measurements to calculate the length, width or area of each rectangle.

a

Length	Width	Area
10.5 cm	8 cm	
	6.5 cm	58.5 cm^2
5.25 cm		15.75 cm^2

b

Length	Width	Area
	4.25 cm	46.75 cm^2
12.75 cm	7.5 cm	
15.5 cm		96.875 cm^2

6 Calculate the area of each triangle.

a

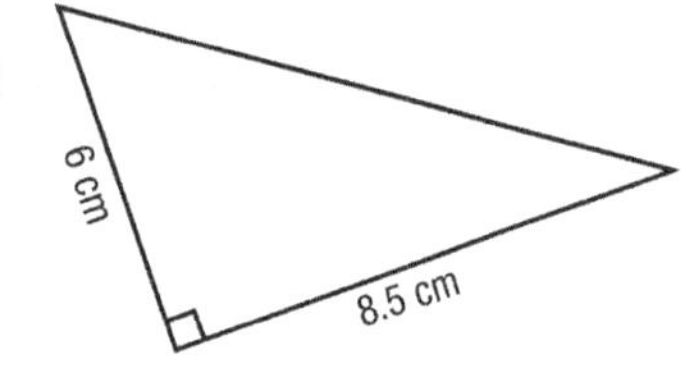

b

11.5 cm
11 cm

c

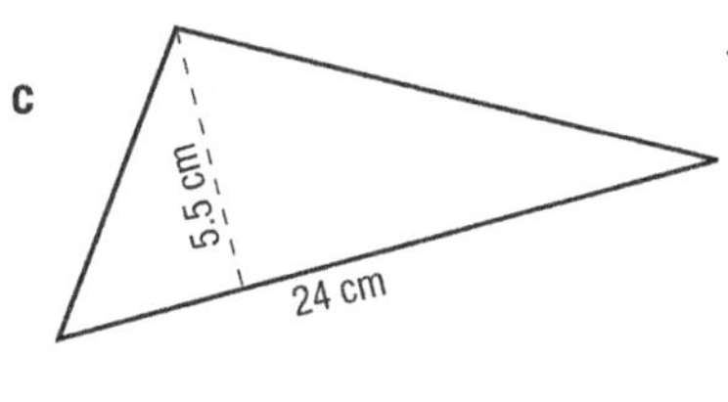

7 Calculate the area of each kite.

a

4 cm
5 cm
12 cm

b

6 cm
6 cm
15 cm

c

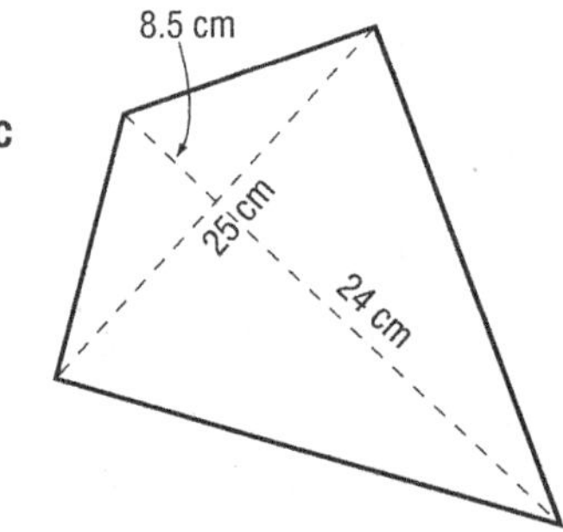

8 Calculate the area of each parallelogram.

a
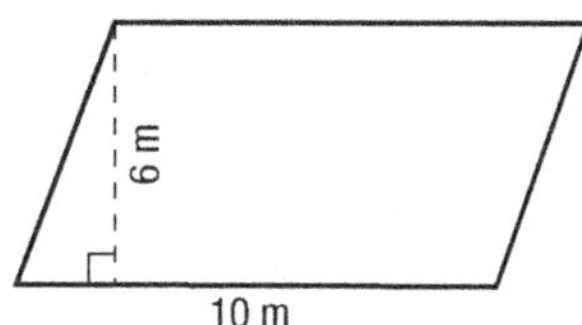

b
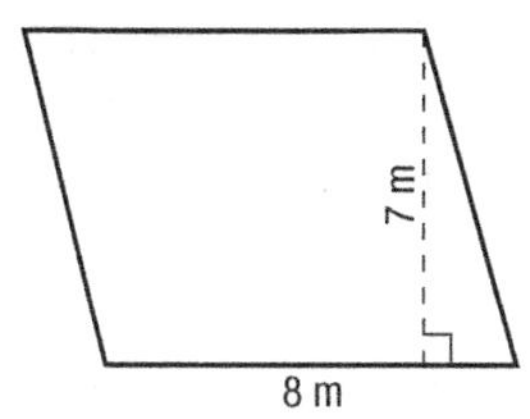

c
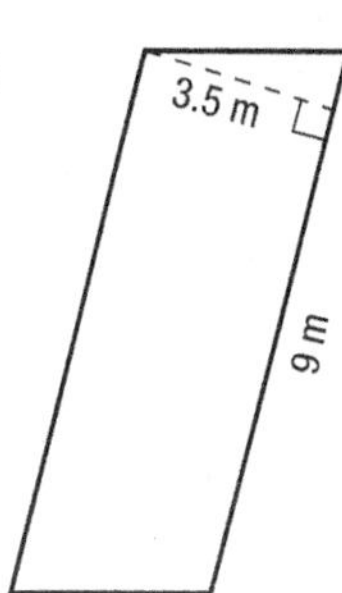

9 Calculate the area of each trapezium.

a
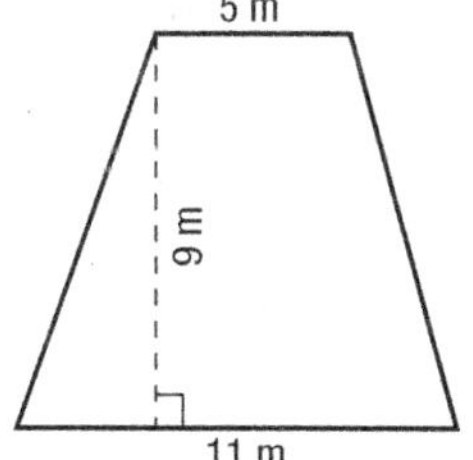

b
10 m
8 m
22 m

c
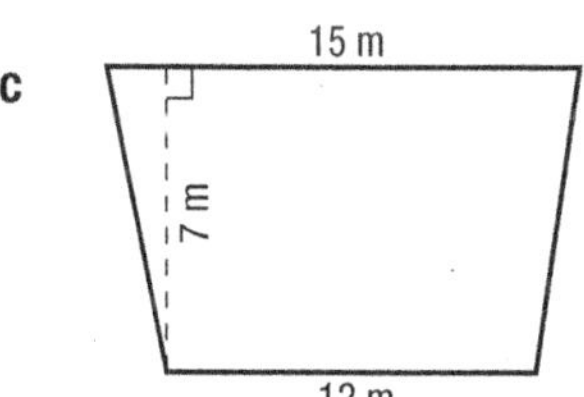

10 Calculate the shaded area of each shape.

a
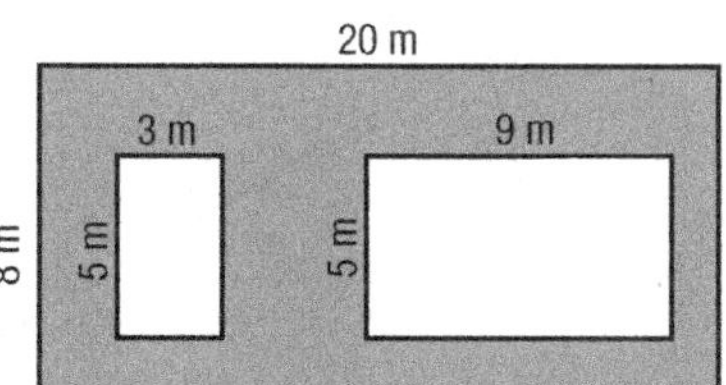

b
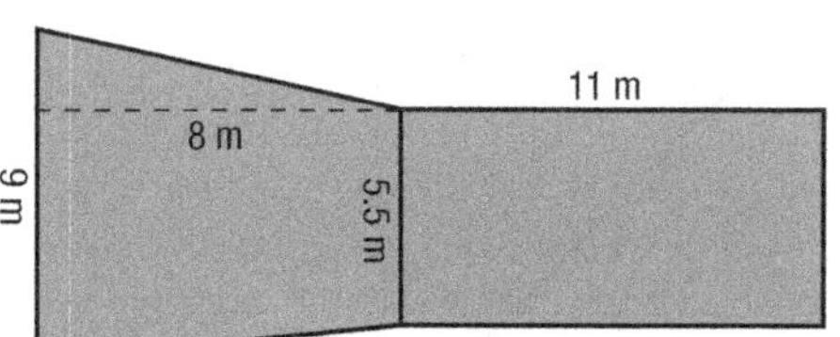

11 Calculate the volume of rectangular prisms with the following dimensions.

a 8 cm high, 13 cm long and 4 cm wide

b 5 cm wide, 11 cm long and 6 cm high

c a cube with 8 m sides

d 15 m long, 7 m wide and 10 m high

12 Calculate the volume of each solid.

a
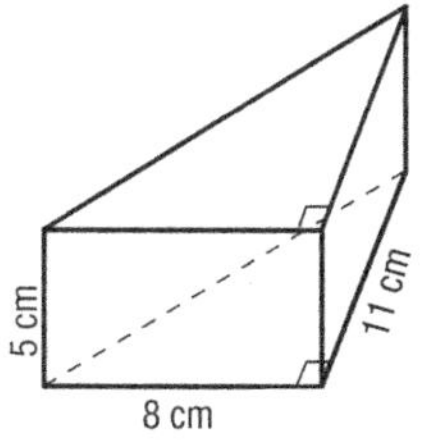

b
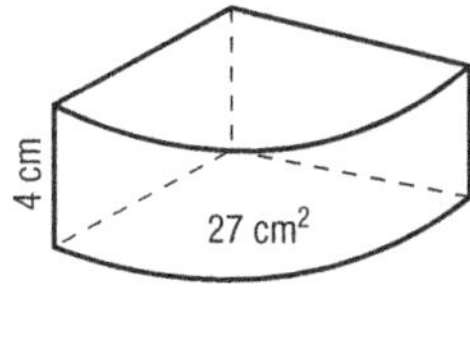

c
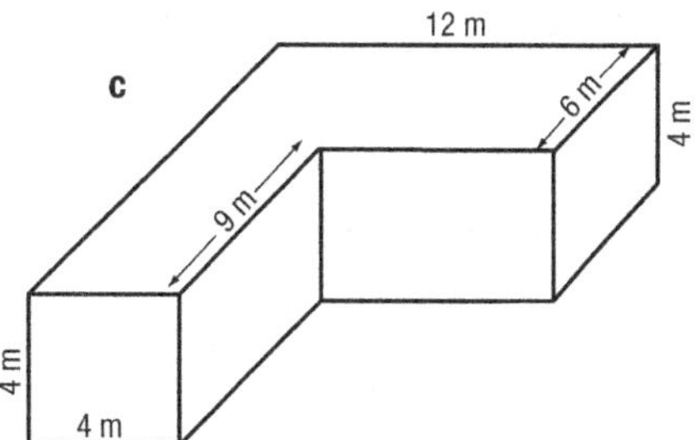

13 What would be the volume of containers with the following capacity?

a 350 mL b 97 mL c 2000 L d 6500 L e 8.5 L f 3.75 L

14 What would be the capacity of containers with the following volume?

a 30 cm^3 b 2400 cm^3 c 4 m^3 d 12.5 m^3 e 915 cm^3 f 7.75 m^3

15 Calculate the unknown angle in each shape.

a
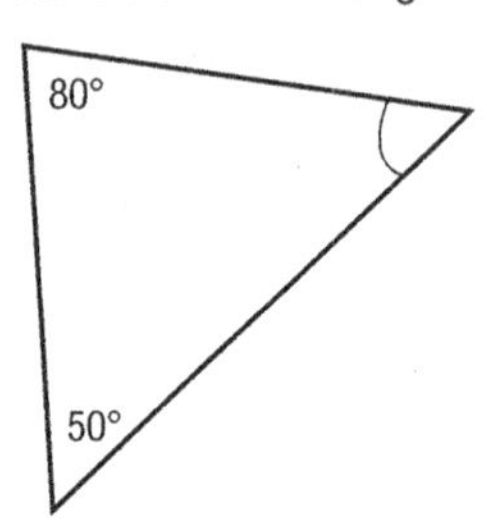

b
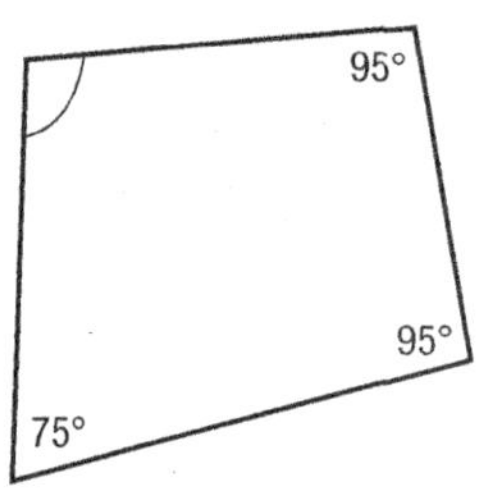

c
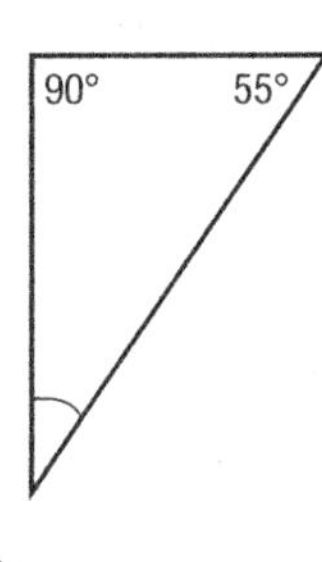

d
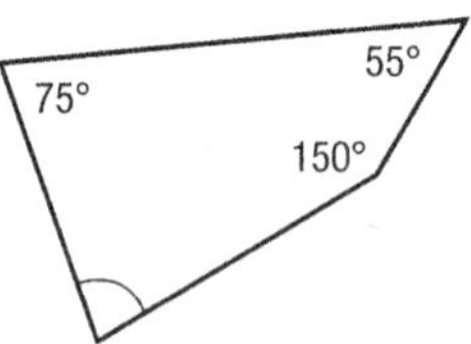

e
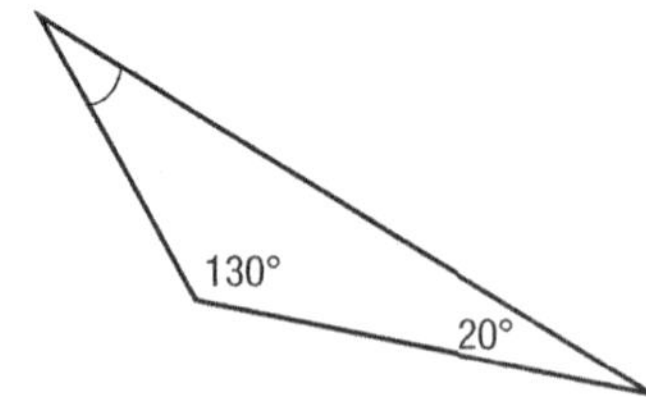

f
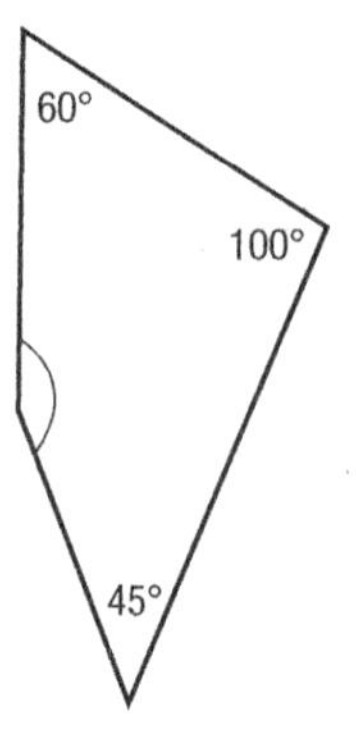

16 Calculate the exterior angle that is marked on each triangle.

a
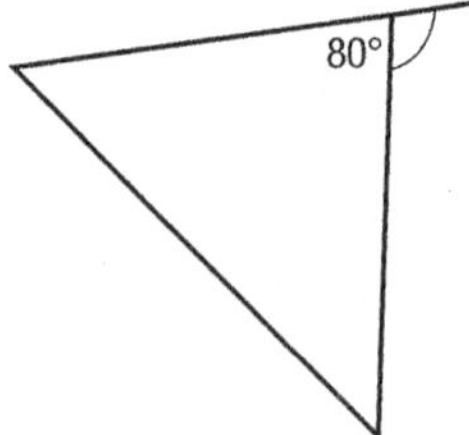

b
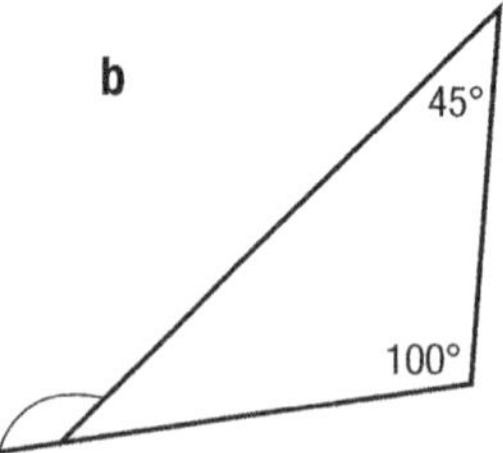

c
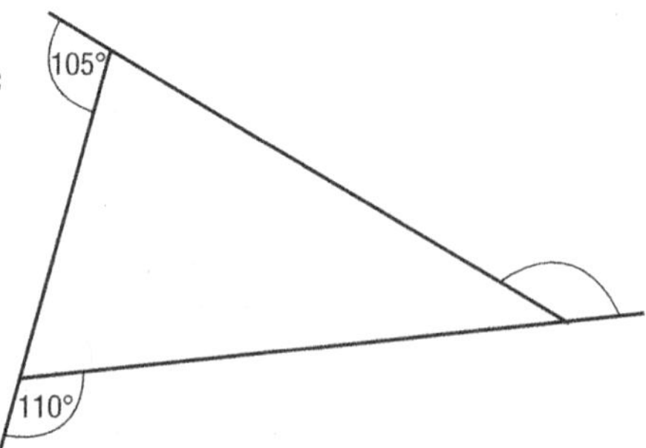

17 Calculate the unknown angle in each diagram.

a
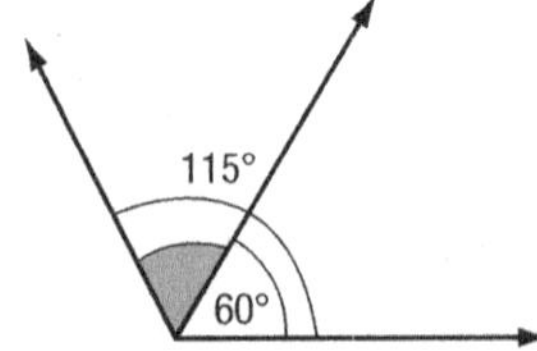

b
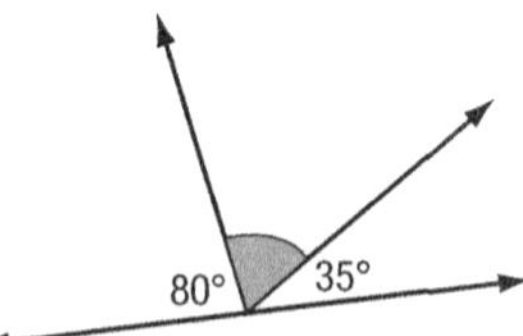

c
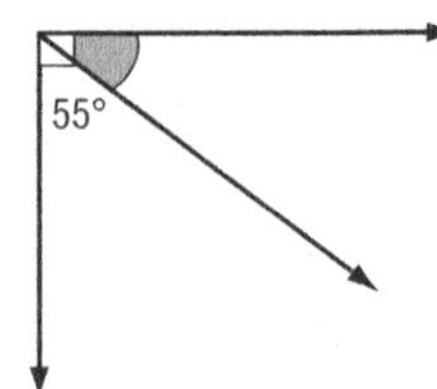

d
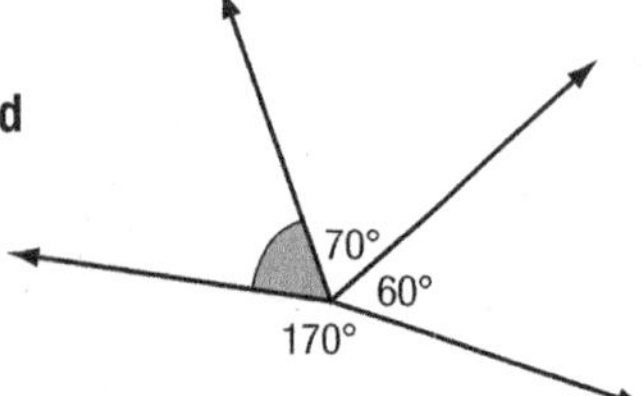

e
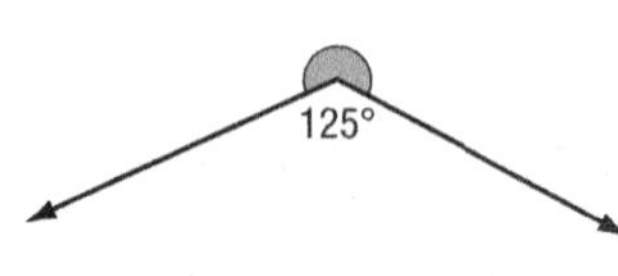

f
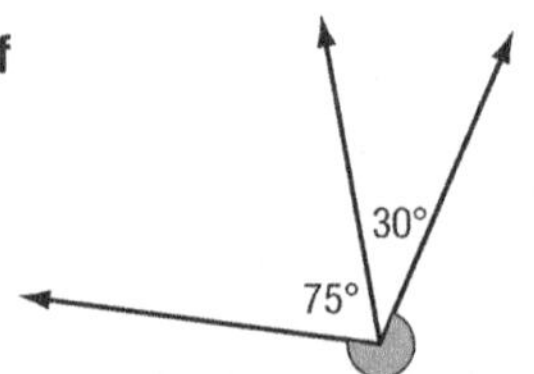

18 What is the bearing of point A in each diagram?

a
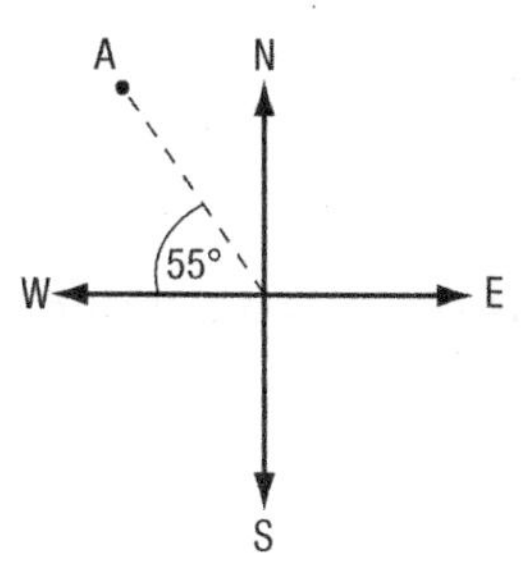

b
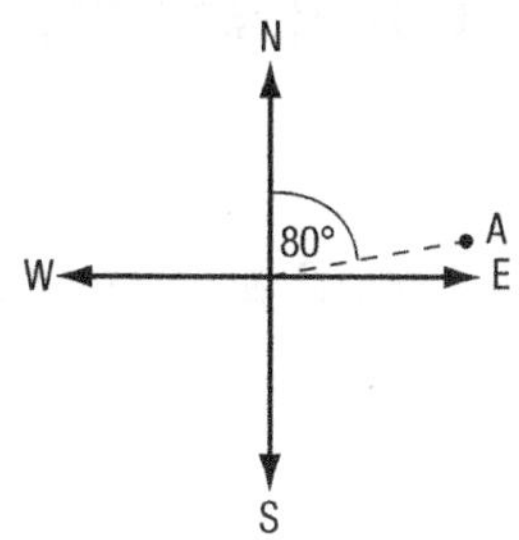

c
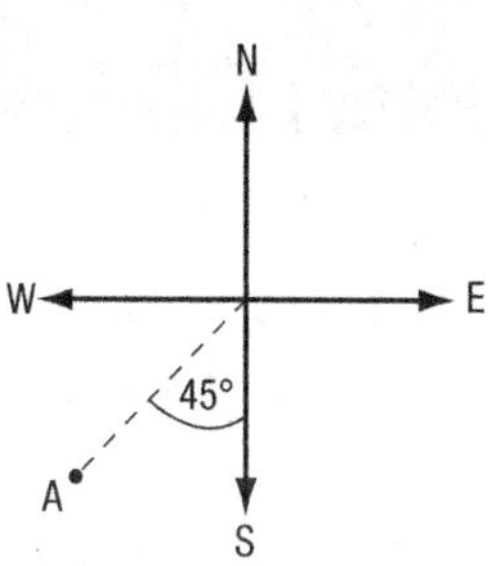

19 Write each bearing as a direction.

a 140° **b** 075° **c** 225° **d** 340° **e** 195° **f** 040°

20 Write each scale as a ratio.

a 1 cm = 2 km **b** 2 cm = 10 m **c** 1 cm = 500 m

21 The scale used on a map is 1:25 000. Calculate the real distance for these distances measured from the map.

a 3 cm **b** 8 cm **c** 2.5 cm **d** 15 cm

22 If the scale used on a map is 1:500 000, how many centimetres would represent these actual distances?

a 10 km **b** 2.5 km **c** 35 km **d** 22.5 km

23 Give ordered pairs for these points on the number plane.

a Point A **b** Point B
c Point C **d** Point D
e Point E **f** Point F

24 What shapes are at the following points?

a (3, –5) **b** (–2, 3)
c (–1, –4) **d** (2, –3)
e (–1, 5) **f** (4, 5)

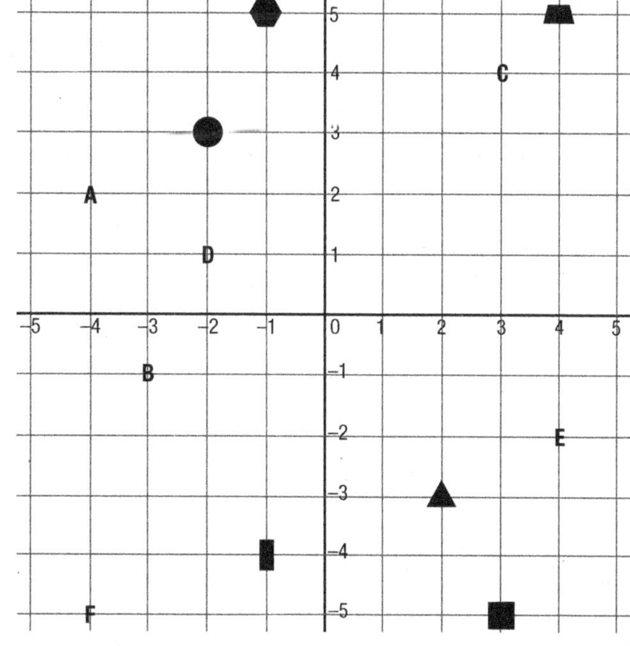

Strand Measurement

7.3.1 Use charts and graphs to read and record weights and solve problems related to weight

Solve weight problems

1 Convert the different weights to the same unit of measurement and calculate.

a $280\text{ g} + 1\frac{1}{2}\text{ kg} + 2.35\text{ kg}$
b $1.2\text{ kg} + 455\text{ g} + 0.75\text{ kg}$
c $2\frac{1}{4}\text{ kg} + 1165\text{ g} + 1.47\text{ kg}$
d $860\text{ kg} + 3.5\text{ t} + 0.7\text{ t}$
e $5.23\text{ t} - 687\text{ kg}$
f $6.075\text{ kg} - 269\text{ g}$
g $45\text{ g} \times 7 + 2\text{ kg }125\text{ g}$
h $36\text{ kg} \times 8 - 712\text{ g}$
i $264\text{ g} \times 4 - 0.86\text{ kg}$

2 Copy and complete the following chart.

Total weight	Number of items of the same weight	Weight of one item
2.184 kg	8	
2.445 t	5	
	7	654 g
1.485 t	9	
	4	1.27 kg
	11	3.927 kg
4.776 kg	6	

3 Write answers to the following in tonnes.

a the total weight of 50 pigs each weighing 78 kg
b the total weight of 75 sacks of rice each weighing 20 kg
c the total weight of 60 bags of potatoes each weighing 26 kg
d the total weight of 45 sacks of coffee beans each weighing 35 kg

4 Write answers to the following in kilograms.

a the total weight of 75 chickens each weighing 320 g
b the total weight of 12 piglets each weighing 4.75 kg
c the total weight of 28 rabbits each weighing 1450 g
d the total weight of 40 ducklings each weighing 550 g

5 The total weight of 5 pigs is 365 kg. What is the average weight of each pig?

6 A farmer loaded 15 bags of grain, each weighing 20 kg, and 12 bags of seed, each weighing 25 kg, onto his truck. What was the total weight of grain and seed that the farmer loaded?

7 If Raymond uses 225 g of food each day to feed his duckling, how many days will a bag containing 3.15 kg of food last?

7.3.4 Manipulate time to solve problems

Read and record time to the nearest minute

Write the time from each clock face in analogue and digital form. For example, the first clock shows 21 to 5 or 4:39.

Convert between time units

1 Convert these units of time.

a	6 weeks to days	**b**	3 hours to minutes	**c**	7 years to months
d	$4\frac{1}{2}$ centuries to years	**e**	11 minutes to seconds	**f**	5 decades to years
g	$5\frac{1}{4}$ hours to minutes	**h**	$3\frac{1}{2}$ years to months	**i**	$4\frac{1}{2}$ minutes to seconds
j	3 days to hours	**k**	$8\frac{1}{2}$ decades to years	**l**	$7\frac{3}{4}$ hours to minutes
m	12 weeks to days	**n**	$6\frac{1}{4}$ years to months	**o**	$9\frac{1}{2}$ days to hours
p	$4\frac{1}{2}$ decades to years	**q**	8 weeks to days	**r**	$2\frac{1}{4}$ centuries to years
s	$2\frac{1}{3}$ hours to minutes	**t**	$7\frac{1}{4}$ minutes to seconds	**u**	$8\frac{2}{3}$ years to months
v	$3\frac{1}{5}$ decades to years	**w**	$7\frac{2}{3}$ hours to minutes	**x**	$5\frac{3}{4}$ centuries to years

2 Convert these minutes to hours, using fractions where necessary.

a	360 minutes	**b**	150 minutes	**c**	270 minutes	**d**	600 minutes	**e**	420 minutes
f	90 minutes	**g**	315 minutes	**h**	105 minutes	**i**	135 minutes	**j**	540 minutes

3 Convert these hours to days, using fractions where necessary.

a	72 hours	**b**	240 hours	**c**	144 hours	**d**	120 hours	**e**	36 hours
f	132 hours	**g**	204 hours	**h**	276 hours	**i**	84 hours	**j**	168 hours

4 Convert these months to years, using fractions where necessary.

a	48 months	**b**	84 months	**c**	18 months	**d**	54 months	**e**	114 months
f	66 months	**g**	138 months	**h**	15 months	**i**	45 months	**j**	87 months

5 Convert these seconds to minutes, using fractions where necessary.

a	240 seconds	**b**	210 seconds	**c**	390 seconds	**d**	105 seconds	**e**	255 seconds
f	435 seconds	**g**	555 seconds	**h**	630 seconds	**i**	525 seconds	**j**	750 seconds

6 Convert these years to centuries, using fractions where necessary.

a	600 years	**b**	1000 years	**c**	450 years	**d**	150 years	**e**	325 years
f	975 years	**g**	1200 years	**h**	5000 years	**i**	675 years	**j**	550 years

7 Convert these years to decades, using fractions where necessary.

a	80 years	**b**	130 years	**c**	45 years	**d**	15 years	**e**	62 years
f	96 years	**g**	31 years	**h**	177 years	**i**	215 years	**j**	75 years

Convert between 12-hour and 24-hour clock time

Remember

In 24-hour time, midday or noon is written as 1200 and midnight is written as 0000 or 2400. The letters a.m. are used to show the period of time from midnight to noon (morning) and the letters p.m. are used to show the period of time from noon to midnight (afternoon and evening).

1 Write these 24-hour clock times as 12-hour clock times, for example: 1840 = 6:40 p.m.

a 1650 **b** 1125 **c** 1345 **d** 2005 **e** 0900 **f** 1400

g 0736 **h** 1727 **i** 1408 **j** 0530 **k** 0055 **l** 1521

m 1934 **n** 0254 **o** 2241 **p** 2359 **q** 0615 **r** 1058

s 1528 **t** 0412 **u** 1207 **v** 1819 **w** 2152 **x** 0026

2 Write these 12-hour clock times as 24-hour clock times.

a 4:15 p.m. **b** 2:35 p.m. **c** 9:23 a.m. **d** 6:12 a.m. **e** 5 p.m. **f** 10:45 a.m.

g 11:08 p.m. **h** 7:10 p.m. **i** 7:48 a.m. **j** 9:55 a.m. **k** 12:05 p.m. **l** 12:20 a.m.

m 1:18 p.m. **n** 3:47 p.m. **o** 5:26 a.m. **p** 8:51 p.m. **q** 4:39 a.m. **r** 2:17 a.m.

s 9:04 p.m. **t** 6:32 p.m.

3 Write the latest time in each group of times.

a 3:15 p.m. 1518 1506 3:13 p.m. 3:15 a.m.

b 2035 10:45 a.m. 2003 10:40 p.m. 1040

c 6:55 a.m. 0658 6 a.m. 0650 6:15 a.m.

d 1340 1:30 p.m. 1350 1:50 a.m. 0159

e 1746 0548 5:45 p.m. 5:40 p.m. 1754

f 2205 10:15 a.m. 2230 10:24 p.m. 2200

g 4:10 p.m. 1604 1620 0445 4:50 a.m.

h 1850 6:15 p.m. 1805 6:35 a.m. 6:30 p.m.

i 2117 6:20 p.m. 5:30 a.m. 0918 10 p.m.

j 1918 7:15 p.m. 7 p.m. 0735 7:05 a.m.

k 2:20 p.m 1400 1425 2:15 p.m. 0250

l 11:10 a.m. 2300 11:40 a.m. 11:15 p.m. 1145

m 1734 5:37 p.m. 0540 5:30 a.m. 1727

n 1:28 a.m. 0145 1312 1:45 p.m. 1356

4

1650	1500	0450	2045	1123
2237	0735	1004	1355	0047

a Which times are between noon and midnight?

b Which times are between 10 a.m and 2 p.m.?

c Which time is closest to 5 p.m.?

d Which time is closest to 7 a.m.?

e Which times are between midnight and 8 a.m.?

f Which times are between 2:30 p.m. and 5 p.m.?

g Which times are between 7 a.m. and 11 a.m.?

h Put the times in order from earliest to latest.

Calculate with time

1 If each film goes for 1 hour and 20 minutes, how much longer will films that have been running for the following times take to complete?

a	56 minutes	b	27 minutes	c	quarter of an hour	d	72 minutes
e	67 minutes	f	38 minutes	g	half an hour	h	19 minutes

2 Students are allowed 2.5 hours to complete an end of year exam. How much longer do these students have left to complete their exam?

	Name	Time Spent
a	Okame	55 minutes
b	Herla	38 minutes
c	Jack	1 hour 35 minutes
d	Kathy	89 minutes
e	Nina	2 hours 12 minutes
f	Felix	76 minutes
g	David	125 minutes
h	Alice	96 minutes

3 These are the departure times for various flights. If each flight takes $2\frac{1}{4}$ hours, what time will they arrive?

a	1020	b	1650	c	0910	d	2115	e	1345	f	0725
g	1755	h	1105	i	1535	j	2340	k	0830	l	1900

4 If each person went to sleep and woke up at the times shown, how long did they sleep for?

	Name	Went to sleep	Woke up
a	Charlie	7:30 p.m.	0600
b	Eileen	2130	8 a.m.
c	Emily	8:45 p.m.	0715
d	Kennedy	2150	7:25 a.m.
e	Sammy	2235	8:05 a.m.
f	Sarah	11:20 p.m.	0640

5 Calculate how long it is between the following pairs of times.

a	5:40 a.m. and 10:17 a.m.	b	9:12 a.m. and 4:47 p.m.	c	7:23 a.m. and 12:07 p.m.
d	8:56 a.m. and 6:33 p.m.	e	0621 and 1347	f	1039 and 1748
g	1445 and 2336	h	1153 and 1914	i	8:12 a.m. and 1647
j	0923 and 3:34 p.m.	k	1809 and 11:48 p.m.	l	12:35 p.m. and 2216
m	6:57 a.m. and 1109	n	0823 and 2:41 p.m.	o	0553 and 1:15 p.m.

Calculate time before and after a given time

Calculate the times under each clock.

1

a $\frac{3}{4}$ hour later
b 2.5 hours earlier
c 35 minutes later

2

a 20 minutes earlier
b $1\frac{1}{4}$ hours earlier
c 50 minutes later

3

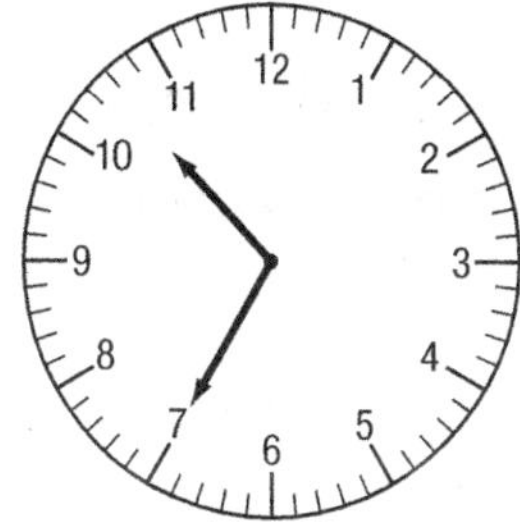

a 1.5 hours earlier
b 40 minutes later
c 1 hour 20 minutes earlier

4

a $1\frac{3}{4}$ hours later
b 85 minutes earlier
c $\frac{3}{4}$ hour earlier

5

a 35 minutes earlier
b 3.25 hours later
c $\frac{1}{2}$ hour earlier

6

a 70 minutes later
b 40 minutes earlier
c 100 minutes later

7

0738

a 2 hours 10 minutes later
b 55 minutes earlier
c $1\frac{1}{4}$ hours earlier

8

2316

a 1.5 hours later
b 50 minutes later
c $\frac{3}{4}$ hour earlier

9

1447

a 2.5 hours earlier
b 45 minutes later
c $1\frac{3}{4}$ hours earlier

Solve time word problems

1 It takes Alex 12 minutes and 27 seconds to walk from school to the trade store. It takes Belinda 10 minutes and 48 seconds to walk the same distance. By how much time is Belinda faster than Alex?

2 Mary bakes eight cakes in 5 hours and 36 minutes.

- **a** What is the average time it takes to bake one cake?
- **b** Use the average time for baking one cake to calculate the time it would take to bake twelve cakes.

3 Brian spent 1 hour and 20 minutes at the pool on Monday, 56 minutes on Tuesday, 83 minutes on Wednesday and $1\frac{3}{4}$ hours on Thursday.

- **a** How much time did Brian spend at the pool over the four days?
- **b** How much more time did he spend at the pool on Thursday than on Tuesday?
- **c** How much less time than 6 hours did Brian spend at the pool?

4 A tennis player spent a total of 10 hours and 9 minutes on court during a tennis tournament. If she played seven matches during the tournament, what was the average time per match?

5 A university student attends lectures for $5\frac{1}{4}$ hours per day from Monday to Friday.

- **a** How many hours per week does the student attend lectures?
- **b** How much time would the student spend at lectures over an eight-week period?

6 Ben watches a 23-minute television program every day of the week. How much television does Ben watch in:

a 1 week? **b** 4 weeks? **c** April? **d** October?

7 Irene arrived at a shopping centre at 9:40 a.m. and spent 3 hours and 35 minutes there. On the way home, she stopped at a market for 85 minutes. It then took her another 17 minutes to get home.

- **a** What time did she leave the shopping centre?
- **b** What time did she arrive home?
- **c** What time did she leave the market?

8 A lecturer earns K8.50 per hour. How much will the lecturer earn if she works the following hours?

a 3 hours **b** 8.5 hours **c** 15 hours **d** 11.5 hours

9 Alice can swim 200 metres in 2 minutes and 11 seconds. John can swim the same distance in 2 minutes and 3 seconds, Isaac can swim it in 1 minute and 58 seconds and Naomi can swim it in 2 minutes and 14 seconds.

- **a** How much faster can Isaac swim 200 metres than Alice?
- **b** If the four swimmers each swam 200 metres as part of an 800-metre relay, what would their time be for the relay?

Assessment Measurement

1 Convert the different weights to the same unit and calculate.

a $2\frac{1}{2}$ kg + 455 g + 1.72 kg

b 0.8 t + 730 kg + 2.56 t

c 7.1 kg – 838 g

d 1.02 t – 579 g

2 Write answers to these in kilograms.

a the total weight of 32 chickens each weighing 330 g

b the total weight of 27 ducklings each weighing 545 g

3 Write the time from each clock face in analogue and digital form.

a 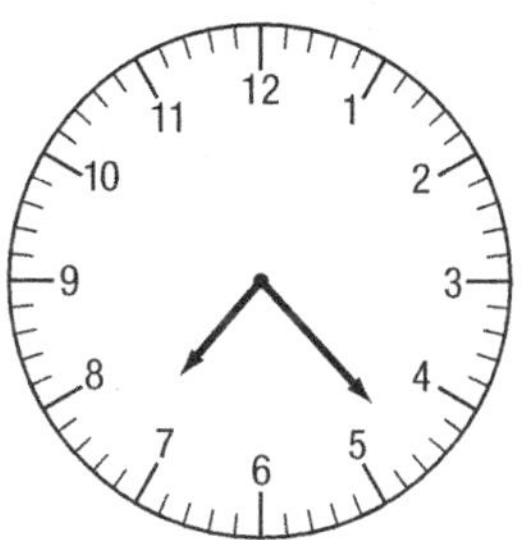

b 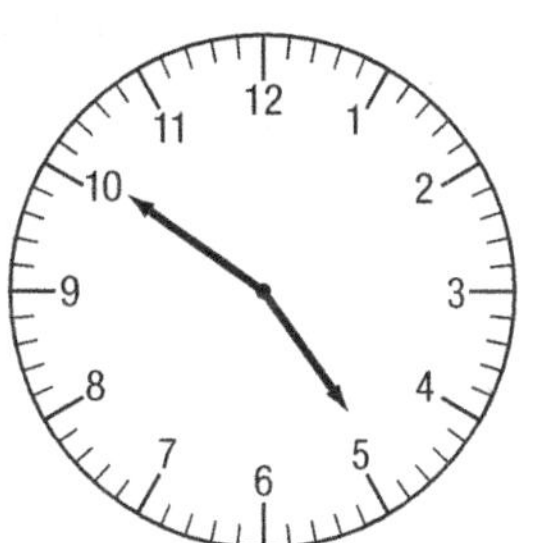

c 

4 Convert these units of time.

a 240 minutes to hours

b $5\frac{1}{2}$ years to months

c 65 years to decades

d $4\frac{2}{3}$ hours to minutes

e 9 weeks to days

f 475 years to centuries

g $7\frac{3}{4}$ minutes to seconds

h 78 months to years

i 11 days to hours

5 Write these 24-hour clock times as 12-hour clock times.

a 1455 b 2017 c 0328 d 1006 e 1842 f 2234

6 Write these 12-hour clock times as 24-hour clock times.

a 7:46 p.m. b 11:29 a.m. c 1:12 p.m. d 2:51 a.m.

e 9:53 p.m. f 5:31 p.m. g 9:05 a.m. h 3:24 p.m.

7 Put these sets of times in order from earliest to latest.

a 2:17 p.m. 1436 1410 2:40 a.m. 1350

b 1735 5:30 p.m. 1748 5:50 a.m. 1640

c 1055 10:15 p.m. 10:50 a.m. 10:15 a.m. 1010

d 2100 8:45 p.m. 0820 8:10 a.m. 2035

e 6:25 a.m. 1815 6 p.m. 1745 0615

8 These are the arrival times for various flights. If each flight takes $1\frac{3}{4}$ hours, what time did they depart?

a 1430 **b** 0915 **c** 1950 **d** 1640 **e** 1105 **f** 1325

9 A film runs for 47 minutes. If the film starts at the following times, what time will it finish?

a 10:25 a.m. **b** 1545 **c** 0920 **d** 1:35 p.m.
e 1650 **f** 11:30 a.m. **g** 2:40 p.m. **h** 2015

10 Calculate how long it is between the following pairs of times.

a 6:50 a.m. and 1210 **b** 2310 and 3:48 a.m. **c** 1608 and 11:20 p.m.
d 2:36 p.m. and 2145 **e** 9:40 a.m. and 1508 **f** 1335 and 9:23 p.m.

11 Copy and complete this chart.

	Current time	$\frac{3}{4}$ hour earlier	1.5 hours later	35 minutes later
a	1924			
b	 [clock]			
c	0437			
d	 [clock]			
e	1308			
f	 [clock]			

Strand Chance and Data

7.4.1 Compare sets of data

Find the mean, median, mode and range

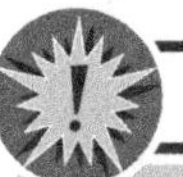

Remember

The **mean** is found by adding a set of scores together and dividing by the number of scores.

The **median** is the middle score or number of a set of scores or numbers. It is found by putting the numbers in order. If there is an even number of data, the median is halfway between the two middle scores.

The **mode** is the score or number that occurs most often in a set of scores or numbers.

The **range** is the spread between the highest and lowest scores and is found by subtracting the lowest score from the highest score.

1 Find the mean of each set of numbers.

a 54, 35, 27, 61, 38
b 16, 75, 66, 34, 29
c 47, 39, 56, 85, 72
d 92, 27, 53, 18, 68, 77
e 63, 49, 75, 84, 71, 58
f 33, 98, 81, 67, 46, 80
g 8.5, 7.2, 9.4, 8.6, 7.8
h 5.8, 3.7, 9.1, 10.6, 9.3
i 8.4, 6.9, 11.5, 5.8, 12.9

2 Find the mean of all:

a numbers between 0 and 20
b even numbers between 0 and 20
c odd numbers between 0 and 20
d numbers between 5 and 25
e even numbers between 5 and 25
f odd numbers between 5 and 25

3 Find the median of each set of numbers.

a 36, 33, 47, 31, 28, 45, 23, 41, 39
b 72, 65, 60, 84, 77, 89, 73, 92, 57
c 28, 17, 36, 33, 15, 29, 33, 19, 26
d 64, 57, 69, 72, 60, 55, 51, 68, 79
e 46, 38, 67, 33, 20, 35, 48, 45, 62, 40
f 93, 87, 90, 99, 82, 106, 88, 80, 73, 91

4 Find the mode of each set of numbers.

a 12, 16, 10, 17, 16, 11, 16, 12, 15, 16
b 9, 6, 2, 7, 6, 3, 5, 6, 7, 6, 2, 9, 7, 8, 6
c $6\frac{1}{2}, 4\frac{1}{2}, 7, 7, 5\frac{1}{2}, 7\frac{1}{2}, 7, 8, 6\frac{1}{2}, 5, 5\frac{1}{2}$
d 9.2, 8.5, 9.6, 8.5, 7.5, 9.2, 9.1, 8.5, 8.8
e 3.4, 2.7, 2.1, 3.8, 3.4, 2.3, 3.4, 2.1, 2.3
f $8\frac{1}{2}, 8, 8\frac{1}{2}, 7\frac{1}{2}, 8, 7, 7\frac{1}{2}, 7\frac{1}{2}, 7, 8\frac{1}{2}, 8\frac{1}{2}$

5 Find the range of each set of numbers.

a 263, 198, 240, 237, 212, 185, 227, 190
b 572, 560, 537, 548, 581, 575, 539, 562
c 87.5, 64.9, 72.8, 75.1, 63.8, 82.7, 61.4
d 56.5, 33.8, 51.2, 49.7, 32.6, 29.8, 54.5
e $3\frac{1}{2}, 7, 8\frac{1}{4}, 7\frac{3}{4}, 6, 5\frac{1}{4}, 9\frac{1}{2}, 9\frac{1}{4}, 8, 4\frac{3}{4}$
f $2\frac{1}{4}, 4\frac{1}{2}, 6, 3\frac{3}{4}, 1\frac{1}{4}, 7\frac{1}{2}, 9, 6\frac{1}{4}, 10, 7\frac{3}{4}$

7.4.3 Calculate probabilities from individual event probabilities

Calculate probability

Help Box

When calculating probability there are two steps.

1. Find the total number of possible outcomes.
2. Write each successful outcome as a fraction of the possible outcomes.

Therefore, the probability of an event = $\frac{\text{number of successful outcomes}}{\text{total number of possible outcomes}}$

Example: To calculate the probability of a coin landing on heads there are two possible outcomes: heads or tails. There is one success (heads) so the probability of a coin landing on heads is $\frac{1}{2}$.

Remember

Probabilities are usually written as fractions in their simplest form. The total of all the probabilities of an event is always 1.

1 Calculate the probability of rolling these numbers on a six-sided dice.

a an even number **b** 3 or 4 **c** 6 **d** a number below 4

2 Calculate the probability of rolling these numbers on a ten-sided dice.

a 5 **b** an odd number **c** 1 or 7 **d** a number above 7

3 A box contains eight cards that are numbered 1 to 8. Calculate the probability of:

a choosing 3 **b** not choosing 1 **c** choosing a number above 4
d not choosing 7 **e** choosing an odd number **f** choosing 2, 5 or 8

4 A container holds six green balls, three red balls and five yellow balls. Calculate the probability of selecting:

a a red ball **b** a green ball **c** a red or yellow ball

5 A purse contains three 10t coins, five 20t coins, two 50t coins and ten K1 coins. Calculate the probability of taking out:

a a 20t coin **b** a 10t coin **c** a K1 coin **d** a 50t coin

6 Tickets numbered from 1 to 50 are placed in a box. Calculate the probability of drawing out:

a an even number **b** a number ending in 5 **c** a number larger than 40
d a multiple of 6 **e** a number between 10 and 20 **f** a number smaller than 15

7 Five girls and seven boys place their names in a hat. One name is drawn out. Calculate the probability of:

- **a** drawing out a girl's name
- **b** not drawing out a boy's name
- **c** drawing out a boy's name
- **d** not drawing out a girl's name

8 A jar contains four red lollies, five green lollies and one purple lolly. Calculate the probability of choosing:

- **a** a red lolly
- **b** a purple lolly
- **c** a green lolly
- **d** a yellow lolly

9 The numbers 6, 31, 19, 28, 15 and 20 are written on cards and put in a box. Calculate the probability of drawing out:

- **a** an even number
- **b** a prime number
- **c** a number above 30
- **d** a number below 10
- **e** a multiple of 5
- **f** a number divisible by 3

10 A standard deck of cards is shuffled and placed face down. Calculate the probability of the top card being:

- **a** a red card
- **b** a black card
- **c** a heart
- **d** a spade
- **e** an ace
- **f** a king
- **g** 9
- **h** a red 5
- **i** a picture card
- **j** 2 or 3
- **k** not a club
- **l** an ace or a jack

11 Calculate the probability of the arrow on the spinner shown here landing on:

- **a** yellow
- **b** black
- **c** red
- **d** white

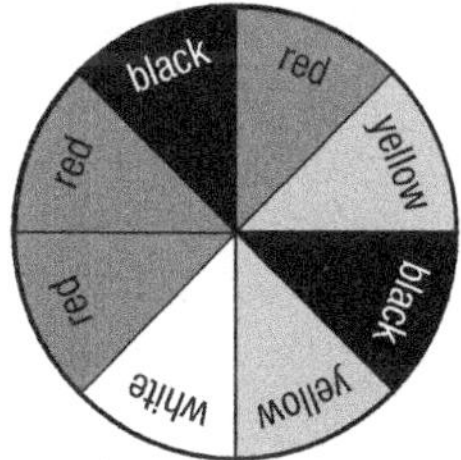

12 Look at the numbers in the box and calculate the probability of selecting:

- **a** an even number
- **b** a number below 100
- **c** a number between 100 and 300
- **d** a number above 400
- **e** a number that is exactly divisible by 3
- **f** a number with 4 tens
- **g** a number that is exactly divisible by 5
- **h** a number with 3 units
- **i** a number between 80 and 220
- **j** a number with 7 tens
- **k** a number that is exactly divisible by 10
- **l** a number below 50

	35		74		12		293		125	
62		240		89		172		109		51
	186		311		533		43		227	
19		450		144		268		153		65

13 If these name cards are turned over and you have to choose one, calculate the probability of picking up:

a	a boy's name	**b**	a girl's name	**c**	a 4-letter name
d	a 5-letter name	**e**	a 6-letter name	**f**	a 7-letter name
g	a name beginning with A	**h**	a name beginning with D	**i**	a name beginning with J
j	a name that contains a 'c'	**k**	a name that has a 'y' on the end	**l**	a name that contains an 'r'

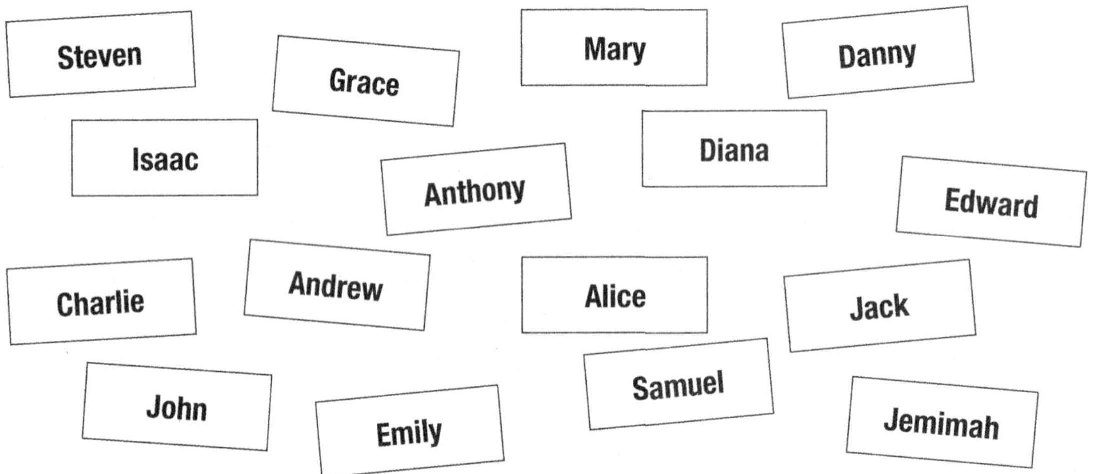

7.4.6 Use a variety of estimation strategies

Estimate with decimals

1 Use estimation to help you place the decimal point in each answer.

a	5.87 × 2.3 = 13501	b	11.7 × 4.8 = 5616	c	9.13 × 5.2 = 47476
d	15.05 × 3.9 = 58695	e	39.9 × 5.2 = 20748	f	26.7 × 7.3 = 19491
g	19.9 × 3.98 = 79202	h	45.2 × 7.19 = 324988	i	11.63 × 4.2 = 48846
j	51.25 × 6.1 = 312625	k	37.88 × 5.8 = 219704	l	28.43 × 7.6 = 216068
m	28.53 × 5.5 = 156915	n	4.77 × 13.2 = 62964	o	2.4 × 0.59 = 1416

2 Look at each multiplication or division and write if the answer will be **larger** or **smaller** than the whole number being multiplied or divided. For example, for 12 × 1.3 the answer will be larger than 12 (the whole number) so you would write 'larger'.

a	15 × 0.7	b	12 ÷ 0.6	c	9 × 1.27	d	10 × 0.92	e	8 ÷ 0.25
f	16 ÷ 2.4	g	25 × 1.05	h	19 × 0.75	i	18 ÷ 0.96	j	14 × 0.38
k	23 ÷ 1.08	l	17 ÷ 0.25	m	15 × 1.63	n	24 × 0.16	o	20 × 1.27
p	19 ÷ 0.75	q	36 ÷ 0.09	r	16 × 3.9	s	8 ÷ 0.33	t	12 ÷ 0.87

3 Estimate the answer to each pair of multiplications or divisions, then write the one in each pair that will result in the **smaller** answer.

a	45 × 1.3	45 × 0.99	b	26 ÷ 0.6	26 ÷ 0.9	c	32 ÷ 0.12	32 ÷ 1.2
d	4 × 5.6	4 × 0.56	e	15 × 0.98	15 × 0.99	f	50 ÷ 1.5	50 ÷ 1.6
g	25 × 0.9	25 × 0.95	h	11 ÷ 1.05	11 ÷ 1.1	i	18 × 0.03	18 × 0.3
j	12 × 1.75	12 × 1.85	k	24 ÷ 1.25	24 ÷ 1.17	l	10 ÷ 0.88	10 ÷ 0.6
m	16 × 1.04	16 × 0.04	n	30 ÷ 1.8	30 ÷ 1.08	o	45 ÷ 0.14	45 ÷ 1.4
p	36 × 0.99	36 × 0.75	q	20 ÷ 1.2	20 ÷ 1.25	r	9 × 1.02	9 × 1.01
s	28 ÷ 0.5	28 ÷ 0.45	t	5 × 0.78	5 × 0.07	u	40 ÷ 0.87	40 ÷ 0.8

4 Use estimation to help you place the decimal point in each answer.

a	10.25 ÷ 4.1 = 25	b	44.1 ÷ 3.6 = 1225	c	1.44 ÷ 0.125 = 1152
d	228.48 ÷ 6.4 = 357	e	75.4 ÷ 1.3 = 58	f	2.87 ÷ 0.35 = 82
g	15.6 ÷ 0.8 = 195	h	1.176 ÷ 0.21 = 56	i	80.66 ÷ 3.7 = 218
j	6.18 ÷ 0.2 = 309	k	8.25 ÷ 0.33 = 25	l	2.52 ÷ 0.5 = 504
m	1.088 ÷ 0.04 = 272	n	36.2 ÷ 2.5 = 1448	o	18.3 ÷ 1.25 = 1464

7.4.7 Estimate sums of money

Estimate with money

1 Copy and complete these charts.

	Number of items and cost per item	Estimate	Cost
a	5 at K9.25 each		
b	4 at K7.75 each		
c	3 at K8.25 each		
d	12 at K1.95 each		
e	6 at K3.80 each		
f	3 at K12.15 each		

	Number of items and cost per item	Estimate	Cost
g	8 at K4.85 each		
h	6 at K7.15 each		
i	9 at K4.35 each		
j	6 at K14.85 each		
k	5 at K17.20 each		
l	8 at K15.95 each		

2 Copy and complete this chart. The first one has been done for you.
For example: round K15.30 to K15.00 and divide by 3 to find the cost of 1 kg (K5.00) then multiply K5.00 by 5 (K25.00) to find an estimated cost for 5 kg. To calculate the exact cost, divide K15.30 by 3 (K5.10) and multiply by 5 (K25.50).

	Amount	Cost	Amount	Estimate	Cost
a	3 kg	K15.30	5 kg	K25.00	K25.50
b	5 kg	K21.00	4 kg		
c	2 kg	K6.80	10 kg		
d	3 kg	K4.95	7 kg		
e	5 kg	K12.00	2 kg		
f	4 kg	K11.80	5 kg		
g	6 kg	K20.70	5 kg		
h	3 kg	K10.65	7 kg		
i	5 kg	K18.75	2 kg		
j	3 kg	K6.90	5 kg		
k	6 kg	K14.70	7 kg		
l	7 kg	K34.65	5 kg		
m	4 kg	K13.80	7 kg		
n	5 kg	K28.75	9 kg		
o	2 kg	K19.80	5 kg		
p	3 kg	K5.25	4 kg		
q	2.5 kg	K6.00	7 kg		
r	4.5 kg	K5.40	5 kg		
s	3.5 kg	K5.25	2 kg		

Assessment Chance and Data

1 Answer the questions below the set of numbers.

27 32 18 28 7 45 12 5

a What is the mean of the set of numbers?

b What is the median of the set of numbers?

c What is the range of the set of numbers?

2 Find the mode of each set of numbers.

a 3.8, 4.1, 2.7, 3.8, 2.5, 4.1, 3.8, 2.9, 2.4, 3.8, 2.7, 2.3

b 106, 98, 112, 94, 110, 94, 100, 91, 94, 100, 106, 115

c $10\frac{1}{2}, 9\frac{1}{4}, 9\frac{1}{4}, 10\frac{3}{4}, 9\frac{1}{4}, 10\frac{1}{2}, 8\frac{3}{4}, 8\frac{1}{2}, 9\frac{1}{4}, 10\frac{1}{2}, 8\frac{3}{4}, 10\frac{1}{2}$

3 Calculate the probability of rolling these numbers on an eight-sided dice.

a an odd number b a number above 2 c 4 or 8 d 5

4 Tickets numbered 26 to 75 are placed in a box. Calculate the probability of drawing out:

a a number larger than 40 b a number between 40 and 50 c a multiple of 7

d a number smaller than 60 e a prime number f a number with 3 tens

g a number with 6 units h a number with a zero on the end

5 Without calculating, write the multiplication or division in each pair that will result in the **larger** answer.

a 27 × 1.3 27 × 1.03 b 16 × 0.97 16 × 0.98 c 70 ÷ 2.08 70 ÷ 2.09

d 36 ÷ 0.5 36 ÷ 1.5 e 52 × 0.75 52 × 0.075 f 23 × 0.8 23 × 0.85

6 Will the answer to the following calculations be larger or smaller than the whole number being multiplied or divided?

a 22 × 0.8 b 71 × 1.09 c 45 × 0.95 d 39 × 1.01

e 54 ÷ 0.7 f 26 ÷ 1.03 g 18 ÷ 1.5 h 67 ÷ 0.02

7 Estimate the total cost of each group of items to the nearest whole kina.

a 5 at K8.80 each b 9 at K6.05 each c 12 at K11.95 each

d 8 at K4.70 each e 6 at K9.15 each f 11 at K15.20 each

Strand **Patterns and Algebra**

7.5.2 Relate number patterns and algebraic statements

Use rules to complete tables

Use the rule to copy and complete these tables.

1 Add 7 to each IN number.

IN	5	15	8	10	23	19
OUT						

2 Subtract 3 from each IN number.

IN	21	16	34	52	40	28
OUT						

3 Multiply each IN number by 4.

IN	11	8	15	22	18	36
OUT						

4 Divide each IN number by 5.

IN	50	35	80	45	25	75
OUT						

5 Halve each IN number and then add 1.

IN	22	38	16	42	50	12
OUT						

6 Add 2 to each IN number and then multiply by 5.

IN	10	7	13	4	23	18
OUT						

7 Subtract 5 from each IN number and then multiply by 3.

IN	16	9	27	14	51	38
OUT						

8 Multiply each IN number by 8 and then add 10.

IN	6	20	9	12	25	32
OUT						

9 Divide each IN number by 3 and then subtract 2.

IN	21	33	60	48	15	54
OUT						

10 Double each IN number and then add 5.

IN	13	17	8	26	53	38
OUT						

11 Add 5 to each IN number and then multiply by 4.

IN	10	17	6	35	26	11
OUT						

12 Subtract 8 from each IN number and then multiply by 10.

IN	16	31	24	57	38	49
OUT						

13 Multiply each IN number by itself and then add 6.

IN	10	8	20	12	9	11
OUT						

14 Divide each IN number by 4 and then add 7.

IN	24	16	40	36	28	60
OUT						

Use and reverse rules to complete tables

Copy and complete these tables by using the rule to find the OUT numbers and reversing the rule to find the IN numbers.

1 Add 4 to each IN number.

IN	15	11	23			
OUT				36	41	25

2 Subtract 8 from each IN number.

IN	27	35	17			
OUT				13	33	52

3 Multiply each IN number by 5.

IN	11	8	15			
OUT				45	90	60

4 Divide each IN number by 7.

IN	56	14	42			
OUT				3	11	9

5 Double each IN number and then add 8.

IN	10	25	16			
OUT				60	48	36

6 Halve each IN number and then subtract 5.

IN	32	46	22			
OUT				14	26	17

7 Add 7 to each IN number and then double.

IN	8	20	17			
OUT				40	64	58

8 Multiply each IN number by 3 and then subtract 1.

IN	11	14	7			
OUT				29	65	89

9 Multiply each IN number by 4 and then add 6.

IN	12		18		9	
OUT		106		62		38

10 Halve each IN number and then add 3.

IN	30		16		42	
OUT		41		15		30

11 Subtract 2 from each IN number and then divide by 4.

IN	30	18			38	62
OUT			11	20		

12 Add 5 to each IN number and then multiply by 4.

IN	20	30			9	23
OUT			80	68		

13 Divide each IN number by 5 and then add 1.

IN	25		60		45	
OUT		11		16		8

14 Double each IN number and then subtract 4.

IN	14		21		17	
OUT		56		84		28

15 Subtract 10 from each IN number and then multiply by 3.

IN	40		23		31	
OUT		24		36		75

16 Multiply each IN number by itself and then add 2.

IN	10		7		20	
OUT		18		146		83

Find and describe number relationships within tables

Work out and record the rule that has been used to complete each table.

1

IN	80	120	50	210	170	360
OUT	8	12	5	21	17	36

2

IN	7	10	5	20	9	15
OUT	56	80	40	160	72	120

3

IN	34	21	86	53	79	45
OUT	22	9	74	41	67	33

4

IN	40	15	37	26	51	19
OUT	65	40	62	51	76	44

5

IN	23	11	32	18	7	27
OUT	69	33	96	54	21	81

6

IN	31	56	29	16	45	50
OUT	24	49	22	9	38	43

7

IN	60	15	35	100	25	10
OUT	13	4	8	21	6	3

8

IN	10	7	2	12	20	18
OUT	98	68	18	118	198	178

9

IN	14	9	22	17	30	27
OUT	29	19	45	35	61	55

10

IN	24	30	14	42	40	28
OUT	11	14	6	20	19	13

11

IN	13	25	10	101	51	6
OUT	36	72	27	300	150	15

12

IN	26	14	38	6	46	32
OUT	14	8	20	4	24	17

13

IN	7	11	4	9	12	10
OUT	49	121	16	81	144	100

14

IN	30	6	22	50	18	4
OUT	20	8	16	30	14	7

15

IN	25	17	31	8	14	36
OUT	49	33	61	15	27	71

16

IN	5	12	30	13	8	20
OUT	26	61	151	66	41	101

17

IN	45	87	32	59	74	98
OUT	56	98	43	70	85	109

18

IN	10	3	20	11	15	7
OUT	150	45	300	165	225	105

Find and use rules to complete tables

Work out and record the rule for each table, then use it to find the missing numbers.

1

IN	6	11	20	9	40	3
OUT	48	88	160			24

2

IN	76	39	53	42	91	67
OUT	56	19		22	71	

3

IN	21	49	34	18	7	26
OUT	71	99			57	

4

IN	10	7	12	5	9	11
OUT	90			45		99

5

IN	67	92	34	71	26	58
OUT	58		25	62		

6

IN	48	60	12	28	72	16
OUT	12	15				4

7

IN	54	33	10	29	72	45
OUT	75		31		93	

8

IN	26	14	38	52		70
OUT	13		19	26	22	

9

IN	6	10	3	16	21	11
OUT	59	99		159		

10

IN	7	12	4	9	15	23
OUT	71	121			151	

11

IN	34		19	45		66
OUT	68	54		90	62	

12

IN	11	20	35	16	5	29
OUT	27	45	75			

13

IN	50	22	42	18	66	30
OUT	24	10		8		

14

IN	15	25	19	40	36	28
OUT	20	40		70		

15

IN	70	12	32	20	8	90
OUT		11		15	9	

16

IN	5		30	8	16	9
OUT	25	121		64		81

17

IN	36	4	64	100	144	
OUT	6		8		12	7

18

IN	7	22	15	2	39	61
OUT	73	223	153			

7.5.3 Substitute numbers for pronumerals

Write and interpret pronumerals in formulae

Remember

A pronumeral is a letter or a symbol that stands for an unknown value.

1 Write each of the rules below as a formula. For example: 'to get m, add 7 to x' becomes $m = x + 7$.

a To get y, subtract 10 from d.
b To get w, add 12 to g.
c To get m, add 5 to s.
d To get a, subtract 8 from b.
e To get c, subtract h from 20.
f To get f, add p to 100.

Remember

Algebra does not use the multiplication sign. For example, instead of $c = d \times 5$, we write $c = 5d$. The number is always written first.

2 Write each of these rules as a formula.

a To get a, multiply b by 7.
b To get h, multiply k by 11.
c To get f, multiply g by 4.
d To get w, multiply y by 5.
e To get d, multiply 10 by c.
f To get n, multiply 8 by p.

3 Use brackets where necessary when writing each of these rules as a formula.

a To get m, first multiply a by 4 and then subtract 2.
b To get x, first add 7 to y and then multiply by 5.
c To get b, first multiply c by 2 and then add 12.
d To get d, first divide f by 4 and then subtract 1.
e To get p, first subtract 3 from q and then divide by 5.
f To get f, first multiply g by 6 and then subtract 10.
g To get s, first divide r by 2 and then multiply by 9.
h To get h, first subtract 9 from j and then multiply by 10.

4 Write the following rules as formulae. To find y:

a Add 9 to x.
b Subtract 15 from x.
c Multiply x by 5 then add 2.
d Add 2 to x then divide by 5.
e Subtract 8 from x then multiply by 3.
f Divide x by 5 then subtract 2.
g Multiply x by 10 then subtract 1.
h Subtract 4 from x then divide by 2.

5 Write these formulae as rules.

a $y = w - 8$
b $x = z + 4$
c $s = 9y$
d $h = 5f$
e $g = 2h - 7$
f $a = 3(b + 8)$
g $d = (c + 5) \div 3$
h $k = 7(m - 3)$
i $p = a \div 3 + 2$
j $r = 6(s - 2)$
k $b = 20 - 3w$
l $v = 8x + 7$

Substitute pronumerals with numerals

Use each formula to find the missing y values. You will need to substitute x in each formula with the given x values.

1 $y = x - 8$

x	23	41	15	76	54	37
y						

2 $y = 4x$

x	8	16	11	6	20	100
y						

3 $y = 40 - x$

x	20	11	4	35	31	17
y						

4 $y = x + 12$

x	42	19	33	67	28	89
y						

5 $y = 2x + 7$

x	9	23	16	31	28	45
y						

6 $y = 4x - 5$

x	11	30	8	15	24	21
y						

7 $y = x^2$

x	10	5	8	12	25	9
y						

8 $y = 2(x + 3)$

x	15	40	13	21	32	27
y						

9 $y = 5(x - 2)$

x	12	5	16	22	8	27
y						

10 $y = 50 - 2x$

x	7	10	21	14	9	12
y						

11 $y = 37 - x \div 2$

x	16	36	28	42	54	30
y						

12 $y = 11 + 5x$

x	5	9	19	12	30	21
y						

13 $y = 3(x \div 4)$

x	40	28	4	100	36	52
y						

14 $y = x^3$

x	2	5	10	3	8	6
y						

15 $y = x \div 3 + 8$

x	27	33	18	9	42	90
y						

16 $y = x^2 - 6$

x	11	7	4	9	12	5
y						

Find the value of pronumerals

1 If $x = 5$, find the value of y in each equation.

a $y = x + 8$ **b** $y = 6x$ **c** $y = x - 4$ **d** $y = 2x - 1$
e $y = 3(x + 3)$ **f** $y = 4x - 9$ **g** $y = 2(x - 3)$ **h** $y = x^2$
i $y = x^2 - 4$ **j** $y = 5x + 7$ **k** $y = 33 - x$ **l** $y = 29 + 2x$

2 If $a = 8$, find the value of b in each equation.

a $b = 5a + 2$ **b** $b = 10a - 25$ **c** $b = a + 11$ **d** $b = 35 - 3a$
e $b = a^2 + 3^2$ **f** $b = 4(a + 3)$ **g** $b = 6(a - 4)$ **h** $b = 6a - 15$
i $b = 4a \div 8$ **j** $b = a^2 - 9$ **k** $b = 17 + 4a$ **l** $b = 3a^2$

3 If $m = 4$ and $p = 6$, find the value of k in each equation.

a $k = m + p$ **b** $k = p - m$ **c** $k = 2m + p$ **d** $k = mp$
e $k = 2mp$ **f** $k = 3m + 4p$ **g** $k = 3p - m$ **h** $k = 4m - 2p$
i $k = 5(m + p)$ **j** $k = m(p - 1)$ **k** $k = 5p + m \div 2$ **l** $k = mp + 9$

4 If $c = 7$ and $d = 9$, find the value of g in each equation.

a $g = cd$ **b** $g = 2c - d$ **c** $g = 4d + c$ **d** $g = 2cd$
e $g = d - c$ **f** $g = c(d + 2)$ **g** $g = d(c - 4)$ **h** $g = cd - 23$
i $g = 5d - 2c$ **j** $g = 19 + d + c$ **k** $g = 22 - d - c$ **l** $g = 4(c + d)$

5 If $w = 3$, $z = 2$ and $v = 5$, find the value of y in each equation.

a $y = w + z + v$ **b** $y = v + z - w$ **c** $y = wz + v$ **d** $y = vz + w$
e $y = wzv$ **f** $y = 2w + 3z + 4v$ **g** $y = w(v - z)$ **h** $y = 5(z + w) + v$
i $y = 100 - wzv$ **j** $y = v^2 + z^2 - w^2$ **k** $y = 4v \div 10 + wz$ **l** $y = vz - 2w$

6 If $s = 10$, $n = 4$ and $h = 2$, find the value of q in each equation.

a $q = 2s - nh$ **b** $q = s - n - h$ **c** $q = 4s + 3n + 5h$ **d** $q = 5(s + n + h)$
e $q = 6s \div 4 + 3nh$ **f** $q = 3snh$ **g** $q = n(s - h)$ **h** $q = 10s - 8n - h$
i $q = 7hn + s$ **j** $q = sh - n$ **k** $q = s^2 - 3nh$ **l** $q = n^2 + (s - h)$

7 Find the value of $9a + 12$ if:

a $a = 6$ **b** $a = 10$ **c** $a = 1$ **d** $a = 11$

8 Find the value of $55 - 4a$ if:

a $a = 9$ **b** $a = 12$ **c** $a = 7.5$ **d** $a = 2\frac{1}{2}$

9 Find the value of $6a - 8$ if:

a $a = 7$ **b** $a = 20$ **c** $a = 5\frac{1}{2}$ **d** $a = 3.5$

Assessment Patterns and Algebra

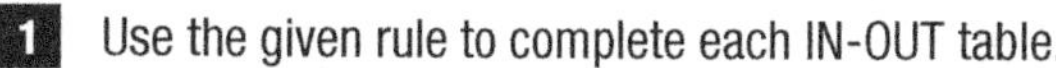

1 Use the given rule to complete each IN-OUT table.

a Halve each IN number and then subtract 3.

IN	40	24	68	110	72	46
OUT						

b Add 1 to each IN number and then multiply by 4.

IN	14	9	30	17	49	26
OUT						

c Divide each IN number by 4 and then add 2.

IN	16		40		28	
OUT		8		4		17

d Double each IN number and then subtract 5.

IN	23		14		36	
OUT		21		39		31

2 Write the rule that has been used to complete each table.

a

IN	5	23	15	9	30	12
OUT	30	138	90	54	180	72

b

IN	21	56	17	39	63	45
OUT	40	75	36	58	82	64

3 Write each of these rules as a formula.

a To get a, add 20 to b.

b To get x, multiply y by 7.

c To get g, subtract h from 18.

d To get w, multiply 4 by z.

e To get c, first multiply d by 8 and then add 2.

f To get k, first subtract 5 from m and then divide by 3.

g To get e, first add 2 to f and then multiply by 9.

h To get p, first divide q by 4 and then multiply by 5.

4 Write these formulae as rules.

a $z = 3w - 1$ **b** $a = 11b$ **c** $g = 2(h + 5)$ **d** $s = 14 - 2t$

5 If $a = 6$, find the value of b in each equation.

a $b = 4a + 5$ **b** $b = 3(a - 1)$ **c** $b = a^2$ **d** $b = 50 - 4a$

6 Find the value of $6p + 9$ if p equals:

a 12 **b** 5 **c** 20 **d** 16 **e** 35 **f** 27

7 Find the value of $4(w - 3)$ if w equals:

a 10 **b** 6 **c** 15 **d** 11 **e** 18 **f** 53

Important Facts

Length and Perimeter

10 millimetres (mm) = 1 centimetre (cm)
100 centimetres (cm) = 1 metre (m)
1000 metres (m) = 1 kilometre (km)
1 mm = $\frac{1}{10}$ cm = 0.1 cm
1 cm = $\frac{1}{100}$ m = 0.01 m
1 m = $\frac{1}{1000}$ km = 0.001 km

Time

60 seconds = 1 minute
60 minutes = 1 hour
24 hours = 1 day
7 days = 1 week
2 weeks = 1 fortnight
52 weeks = 1 year
12 months = 1 year
365 days = 1 year
366 days = 1 leap year
10 years = 1 decade
100 years = 1 century
1000 years = 1 millennium
a.m. = ante meridiem – the period between 12 midnight and 12 noon
p.m. = post meridiem – the period between 12 noon and 12 midnight
12-hour time divides a day into two 12-hour periods: a.m. and p.m.
24-hour time divides a day into twenty-four 1-hour periods, starting at midnight

Capacity

1000 millilitres (mL) = 1 litre (L)
1 mL = $\frac{1}{1000}$ L = 0.001 L

Weight

1000 grams (g) = 1 kilogram (kg)
1000 kilograms (kg) = 1 tonne (t)
1 g = $\frac{1}{1000}$ kg = 0.001 kg
1 kg = $\frac{1}{1000}$ t = 0.001 t

Money

100 toea (t) = 1 kina (K)
1 t = $\frac{1}{100}$ K = K0.01

Area

10 000 square centimetres (cm^2) = 1 square metre (m^2)
10 000 square metres (m^2) = 1 hectare (ha)
100 hectares (ha) = 1 square kilometre (km^2)
Area of a rectangle = Length × Width
Area of a triangle = (Height × Base) ÷ 2
Area of a parallelogram = Base × Height

Temperature

Temperature is usually measured in degrees Celsius (°C). On the Celsius scale, water freezes at 0 degrees and boils at 100 degrees.

Volume

100 000 cubic centimetres (cm^3) = 1 cubic metre (m^3)
Volume of a rectangular prism = Length × Width × Height
Volume of a triangular prism = (Area of base) × Height

Volume and capacity

1 mL = 1 cm^3; 1 L = 1000 cm^3; 1000 L = 1 m^3

Circumference of circles

Circumference can be calculated by either of these formulae: $C = \pi D$ or $C = 2\pi r$ where C is circumference, D is diameter, r is radius and π represents 3.14

Angles

Angles are measured in degrees.

A right angle = 90°

A straight angle = 180°

A full turn = 360°

Acute angles measure between 0° and 90°.

Obtuse angles measure between 90° and 180°.

Reflex angles measure between 180° and 360°.

The interior angles of a triangle total 180°.

The exterior angles of a triangle total 360°.

The interior angles of a quadrilateral total 360°.

Place value

The value of a digit in a number is determined by its place value or position.

Millions	Hundreds of thousands	Tens of thousands	Thousands	Hundreds	Tens	Ones	Tenths	Hundredths	Thousandths

Index numbers

Index numbers (indices) are used to show that a number is multiplied by itself a certain number of times.

Examples:

In 4^3 the index 3 shows that 4 is multiplied by itself 3 times: $4 \times 4 \times 4$.

In 3^5 the index 5 shows that 3 is multiplied by itself 5 times: $5 \times 5 \times 5 \times 5 \times 5$

Directed numbers

Directed numbers are positive and negative whole numbers including zero.

A minus sign is always used to show negative numbers, for example –5, –10 and –127.

Positive numbers can be shown with or without a plus sign, for example +2 or 2, +17 or 17 and +36 or 36.

Order of operations

Do **brackets** first, then **of**, then **multiplication** and **division** in the order they appear from left to right, and finally **addition** and **subtraction** in the order they appear from left to right.

Answers

Strand: Number and Application

Page 2

1 62 735 **2** 13 841 **3** 19 040
4 71 024 **5** 800 397 **6** 309 101
7 480 961 **8** 391 921 **9** 124 158
10 451 429 **11** 471 625 **12** 169 376
13 132 719 **14** 313 819 **15** 217 204

The SUPER TOTAL is 8 058 700 (two times 4 029 350)

Page 3

1 638 389, 619 793, 614 716, 573 897, 415 172, 328 685, 318 909, 244 397, 238 068, 179 631, 152 546, **144 002**

2 347 729, 326 211, 276 215, 269 632, 182 107, 162 538, 155 829, 142 743, 114 258, 81 131, 73 138, **56 879**

3 832 507, 820 474, 783 916, 674 477, 669 310, 589 484, 543 259, 537 481, 324 315, 269 394, 234 837, **229 189**

4 828 172, 772 000, 607 017, 567 511, 565 659, 469 195, 420 604, 187 489, 159 807, 151 331, 139 293, **64 700**

Page 4

1 **a** 10 452 **b** 11 725 **c** 24 310 **d** 28 776
e 38 688 **f** 50 214 **g** 486 594 **h** 165 084
i 291 300 **j** 276 381 **k** 390 176 **l** 197 157
m 171 035 **n** 490 008 **o** 534 375

2 **a** 169 004 **b** 88 332 **c** 292 795 **d** 81 450
e 244 584 **f** 772 048 **g** 569 646 **h** 1 002 364
i 581 031 **j** 853 176 **k** 508 288 **l** 1 434 375
m 1 455 141 **n** 1 012 928 **o** 1 818 978

3 **a** 79 146 **b** 14 262 **c** 47 328 **d** 30 245
e 174 090 **f** 209 344 **g** 233 898 **h** 177 732
i 72 885 **j** 191 550 **k** 97 344 **l** 314 942
m 406 295 **n** 675 104 **o** 1 247 624 **p** 1 358 899

4 **a** and **d**

Page 5

Game

Pages 6–7

1 **a** $556\frac{3}{5}$ **b** $1789\frac{1}{2}$ **c** $421\frac{1}{2}$ **d** $819\frac{1}{6}$
e 829 **f** 5248 **g** $23\,969\frac{1}{3}$ **h** $5868\frac{4}{5}$
i $4326\frac{4}{7}$ **j** $6571\frac{7}{8}$

2 TELEFOMIN

3 AUSTRALIA

4 **a** and **g**; **b** and **e**; **c** and **h**; **d** and **f**

5 **a** $263\frac{3}{5}$ **b** $209\frac{20}{23}$ **c** $208\frac{8}{35}$ **d** $215\frac{7}{12}$
e $167\frac{5}{7}$ **f** $225\frac{5}{6}$ **g** $252\frac{17}{27}$ **h** $223\frac{5}{8}$
i $1621\frac{1}{2}$ **j** $1482\frac{18}{25}$ **k** $2061\frac{10}{11}$ **l** $4013\frac{11}{19}$
m $1520\frac{21}{31}$ **n** $2205\frac{37}{42}$ **o** $2265\frac{1}{2}$ **p** $3190\frac{9}{17}$

6 **a**, **d**, **e** and **h**

7 **b**, **d**, **f** and **h**

8 **a** 415 **b** 284 **c** 693 **d** 316 **e** 578 **f** 467
g 964 **h** 549 **i** 354 **j** 243 **k** 488 **l** 725
m 854 **n** 536 **o** 663 **p** 784

The total is 8913, which is less than 10 000.

Page 8

1 $1719\frac{1}{4}$ **2** 19 068 kg or 19.068 t
3 447 550 L **4** 137 177 L

5 **a** K124 992 **b** K214 272 **c** K428 544 **d** K535 680

6 **a** K1 592 208 **b** K241 242 **c** K634 393

7 K218 985

8 **a** 7 229 184 km^2 **b** 252 195 km^2 **c** 390 119 km^2 **d** 7 962 491 km^2
e 84 910 km^2

9 K58 310.40

Pages 9–10

1 **a** $\frac{11}{20}$ **b** $\frac{1}{3}$ **c** $\frac{1}{2}$ **d** $\frac{7}{12}$ **e** $\frac{3}{5}$ **f** $\frac{4}{30}$
g $\frac{3}{10}$ **h** $\frac{3}{5}$ **i** $\frac{5}{40}$ **j** $\frac{9}{16}$ **k** $\frac{5}{8}$ **l** $\frac{7}{8}$
m $\frac{7}{10}$ **n** $\frac{8}{16}$

2 **a** $\frac{9}{12}$ **b** $\frac{8}{12}$ **c** $\frac{6}{12}$ **d** $\frac{2}{6}$ **e** $\frac{4}{16}$ **f** $\frac{8}{10}$
g $\frac{30}{100}$ **h** $\frac{10}{12}$ **i** $\frac{9}{24}$ **j** $\frac{4}{10}$ **k** $\frac{2}{16}$ **l** $\frac{70}{100}$
m $\frac{12}{27}$ **n** $\frac{15}{35}$ **o** $\frac{15}{24}$

3 **a** $\frac{3}{12}$ **b** $\frac{5}{10}$ **c** $\frac{8}{12}$ **d** $\frac{30}{50}$ **e** $\frac{28}{32}$
f $\frac{9}{12}$ **g** $\frac{6}{36}$ **h** $\frac{35}{50}$ **i** $\frac{8}{64}$ **j** $\frac{5}{25}$
k $\frac{30}{40}$ **l** $\frac{6}{20}$ **m** $\frac{8}{14}$ **n** $\frac{25}{30}$ **o** $\frac{8}{18}$
p $\frac{9}{24}$ **q** $\frac{6}{9}$ **r** $\frac{25}{50}$ **s** $\frac{4}{10}$ **t** $\frac{15}{27}$
u $\frac{9}{21}$ **v** $\frac{42}{48}$ **w** $\frac{25}{35}$ **x** $\frac{33}{36}$ **y** $\frac{20}{25}$

4 **a** $\frac{2}{3}$ **b** $\frac{2}{3}$ **c** $\frac{1}{4}$ **d** $\frac{5}{6}$ **e** $\frac{5}{8}$ **f** $\frac{2}{5}$
g $\frac{1}{5}$ **h** $\frac{4}{7}$ **i** $\frac{3}{5}$ **j** $\frac{4}{7}$ **k** $\frac{2}{3}$ **l** $\frac{5}{9}$
m $\frac{2}{9}$ **n** $\frac{3}{5}$ **o** $\frac{5}{6}$ **p** $\frac{1}{4}$ **q** $\frac{1}{7}$ **r** $\frac{1}{4}$
s $\frac{3}{4}$ **t** $\frac{3}{5}$ **u** $\frac{3}{8}$ **v** $\frac{3}{5}$ **w** $\frac{3}{5}$ **x** $\frac{1}{3}$

Pages 10–11

1 **a** $\frac{10}{3}$ **b** $\frac{12}{5}$ **c** $\frac{7}{4}$ **d** $\frac{9}{2}$ **e** $\frac{23}{8}$ **f** $\frac{11}{6}$
g $\frac{19}{8}$ **h** $\frac{19}{5}$ **i** $\frac{9}{7}$ **j** $\frac{17}{3}$ **k** $\frac{17}{10}$ **l** $\frac{17}{4}$
m $\frac{39}{7}$ **n** $\frac{15}{8}$ **o** $\frac{20}{9}$ **p** $\frac{19}{3}$ **q** $\frac{38}{11}$ **r** $\frac{23}{10}$
s $\frac{15}{2}$ **t** $\frac{25}{6}$ **u** $\frac{29}{8}$ **v** $\frac{35}{4}$ **w** $\frac{8}{5}$ **x** $\frac{31}{3}$
y $\frac{34}{9}$ **z** $\frac{47}{8}$

2 **a** 28 **b** 23 **c** 16 **d** 8 **e** 32 **f** 42
g 7 **h** 15 **i** 42 **j** 17 **k** 22 **l** 22
m 26 **n** 39 **o** 22 **p** 32 **q** 14 **r** 13
s 14 **t** 38 **u** 58 **v** 28 **w** 25 **x** 6

3 a $4\frac{2}{3}$ b $3\frac{1}{3}$ c $4\frac{1}{6}$ d $2\frac{1}{7}$ e $2\frac{3}{8}$ f $7\frac{5}{6}$
g $4\frac{1}{7}$ h $4\frac{5}{12}$ i $6\frac{1}{5}$ j $4\frac{4}{9}$ k $5\frac{1}{4}$ l $4\frac{1}{8}$
m $2\frac{5}{6}$ n $7\frac{1}{5}$ o $6\frac{3}{4}$ p $2\frac{4}{7}$ q $2\frac{2}{5}$ r $6\frac{5}{6}$
s $5\frac{4}{5}$ t $3\frac{2}{7}$ u $2\frac{2}{9}$ v $10\frac{2}{3}$ w $6\frac{1}{4}$ x $3\frac{7}{9}$
y $4\frac{2}{5}$ z $5\frac{3}{7}$

4 a $1\frac{1}{2}$ b $2\frac{3}{4}$ c $4\frac{1}{10}$ d $\frac{10}{7}$ e $\frac{27}{6}$ f $3\frac{2}{3}$
g $\frac{19}{9}$ h $\frac{23}{8}$ i $5\frac{1}{6}$ j $\frac{21}{8}$ k $2\frac{1}{2}$ l $\frac{7}{5}$
m $\frac{33}{7}$ n $5\frac{2}{5}$ o $\frac{25}{4}$ p $1\frac{3}{8}$ q $3\frac{1}{10}$ r $\frac{13}{6}$
s $3\frac{3}{4}$ t $\frac{25}{4}$ u $4\frac{2}{9}$ v $3\frac{5}{11}$ w $\frac{16}{5}$ x $\frac{23}{3}$

Pages 12–13

1 a $1\frac{2}{3}$ b $1\frac{1}{10}$ c $2\frac{3}{5}$ d $3\frac{4}{7}$ e $3\frac{8}{9}$ f $\frac{41}{100}$
g $4\frac{2}{5}$ h $2\frac{1}{12}$ i $3\frac{1}{3}$ j $4\frac{3}{10}$ k $2\frac{3}{10}$ l $4\frac{5}{8}$
m $2\frac{7}{10}$ n $3\frac{1}{2}$ o $2\frac{1}{4}$ p $5\frac{9}{10}$ q $7\frac{1}{8}$ r $3\frac{1}{9}$
s $5\frac{7}{8}$ t $6\frac{3}{4}$ u $6\frac{1}{6}$ v $11\frac{7}{10}$ w $8\frac{1}{15}$ x $3\frac{7}{12}$

2 a $4\frac{7}{10}$ b $8\frac{1}{3}$ c $6\frac{3}{4}$ d $5\frac{1}{5}$ e $7\frac{1}{3}$ f $4\frac{2}{3}$
g $10\frac{7}{8}$ h $6\frac{3}{5}$ i $6\frac{7}{10}$ j $6\frac{3}{4}$ k $11\frac{5}{8}$ l $8\frac{5}{18}$
m $7\frac{5}{12}$ n $11\frac{1}{16}$ o $6\frac{4}{5}$ p $10\frac{3}{4}$ q $8\frac{1}{4}$ r 5
s $8\frac{3}{4}$ t $13\frac{15}{16}$ u $6\frac{13}{21}$ v 10 w 3 x $13\frac{23}{24}$

3 a $3\frac{1}{3}$ b $4\frac{1}{2}$ c $6\frac{1}{9}$ d $8\frac{2}{9}$ e $3\frac{1}{6}$ f $2\frac{8}{21}$
g $11\frac{1}{4}$ h $6\frac{1}{14}$ i $3\frac{4}{33}$ j $5\frac{1}{10}$ k $5\frac{7}{12}$ l $9\frac{5}{12}$
m $12\frac{1}{24}$ n $4\frac{16}{35}$ o $4\frac{4}{45}$ p $11\frac{9}{28}$ q $3\frac{35}{44}$ r $3\frac{23}{24}$

Pages 14–15

1 a $\frac{1}{2}$ b $\frac{5}{12}$ c $\frac{9}{11}$ d $\frac{11}{20}$ e $1\frac{11}{15}$ f $1\frac{1}{2}$
g $2\frac{3}{8}$ h $3\frac{1}{2}$ i $1\frac{2}{3}$ j $\frac{61}{100}$ k $2\frac{5}{6}$ l $1\frac{1}{2}$
m $2\frac{9}{10}$ n $2\frac{11}{12}$ o $2\frac{3}{8}$ p $\frac{7}{9}$ q $2\frac{4}{5}$ r $1\frac{5}{6}$
s $2\frac{13}{15}$ t $2\frac{1}{14}$ u $2\frac{13}{21}$ v $2\frac{7}{10}$ w $2\frac{9}{10}$ x $2\frac{24}{35}$

2 INDONESIA

3 a $\frac{2}{3}$ b $3\frac{2}{21}$ c $1\frac{19}{40}$ d $1\frac{5}{18}$ e $2\frac{5}{14}$ f $2\frac{3}{4}$
g $2\frac{5}{6}$ h $2\frac{2}{3}$ i $1\frac{1}{2}$ j $2\frac{3}{40}$ k $2\frac{1}{24}$ l $1\frac{11}{12}$
m $2\frac{11}{20}$ n $1\frac{7}{18}$ o $4\frac{22}{35}$ p $2\frac{2}{3}$ q $1\frac{32}{55}$ r $1\frac{20}{21}$

Pages 16–17

1 a $\frac{2}{63}$ b $\frac{12}{55}$ c $\frac{7}{15}$ d $\frac{25}{48}$ e $\frac{1}{7}$ f $\frac{3}{5}$
g $\frac{7}{12}$ h $\frac{10}{63}$ i $\frac{3}{20}$ j $\frac{4}{9}$

2 a $4\frac{3}{8}$ b $3\frac{23}{40}$ c $5\frac{5}{12}$ d $6\frac{13}{40}$ e 4 f $6\frac{13}{20}$
g $13\frac{1}{15}$ h $6\frac{13}{63}$ i $9\frac{5}{24}$ j $2\frac{23}{55}$ k $10\frac{2}{3}$ l $5\frac{5}{8}$

3 a $13\frac{1}{3}$ b $22\frac{4}{5}$ c $1\frac{9}{40}$ d $1\frac{7}{16}$ e $7\frac{5}{7}$ f $3\frac{2}{21}$
g $1\frac{23}{40}$ h $3\frac{1}{9}$ i $\frac{19}{24}$ j $23\frac{8}{9}$

4 a $2\frac{17}{20}$ b $\frac{23}{24}$ c 20 d $1\frac{11}{21}$ e 64 f 15
g $1\frac{1}{27}$ h $1\frac{13}{32}$ i $1\frac{25}{63}$ j 49

5 a $32\frac{2}{3}$ b 63 c $1\frac{1}{21}$ d $1\frac{9}{40}$ e 10 f $4\frac{10}{11}$
g $4\frac{11}{16}$ h $\frac{4}{33}$ i $5\frac{45}{56}$ j $1\frac{3}{20}$ k $4\frac{1}{20}$ l $29\frac{1}{3}$
m $\frac{5}{32}$ n $8\frac{3}{40}$ o $7\frac{8}{35}$ p $60\frac{5}{7}$ q $31\frac{1}{9}$ r $\frac{21}{55}$
s $13\frac{4}{5}$ t $\frac{91}{120}$ u $3\frac{3}{5}$

Pages 18–19

1 a $1\frac{1}{2}$ b $3\frac{1}{5}$ c $1\frac{7}{8}$ d $1\frac{7}{18}$ e $\frac{4}{5}$ f $1\frac{11}{24}$
g $3\frac{1}{9}$ h $3\frac{7}{11}$ i $\frac{14}{15}$ j $1\frac{7}{8}$ k $1\frac{13}{22}$ l $2\frac{1}{12}$
m $\frac{2}{3}$ n $1\frac{21}{22}$ o $2\frac{3}{10}$ p $9\frac{13}{35}$ q $2\frac{3}{8}$ r $3\frac{3}{5}$
s $\frac{16}{225}$ t $\frac{7}{20}$

2 a $\frac{7}{40}$ b $10\frac{2}{3}$ c $\frac{2}{5}$ d $1\frac{1}{3}$ e $\frac{38}{63}$ f $12\frac{1}{2}$
g $\frac{8}{27}$ h $1\frac{1}{5}$ i $\frac{8}{99}$ j $2\frac{3}{11}$ k $\frac{7}{16}$ l $6\frac{6}{11}$
m $\frac{29}{30}$ n $\frac{5}{14}$ o $\frac{7}{48}$

3 a $\frac{3}{8}$ b $2\frac{3}{16}$ c $1\frac{47}{70}$ d $4\frac{2}{27}$ e $2\frac{1}{2}$ f $\frac{27}{56}$
g $\frac{102}{217}$ h $2\frac{3}{34}$ i $1\frac{1}{7}$ j $1\frac{1}{20}$ k $1\frac{37}{117}$ l 3

4 a $\frac{29}{40}$ b $4\frac{1}{8}$ c $1\frac{19}{21}$ d $2\frac{10}{11}$ e 5 f $2\frac{1}{17}$
g $\frac{5}{24}$ h $1\frac{47}{52}$ i $11\frac{4}{7}$ j $1\frac{8}{13}$ k $11\frac{21}{25}$ l $\frac{19}{36}$
m $1\frac{79}{161}$ n $\frac{20}{121}$ o $1\frac{46}{119}$ p $\frac{35}{78}$ q $\frac{29}{104}$ r $2\frac{46}{49}$
s $4\frac{17}{22}$ t $9\frac{5}{9}$ u $\frac{63}{128}$

Page 20

1 $14\frac{2}{3}$ cups **2** $6\frac{1}{8}$ litres **3** $1\frac{23}{24}$ kg **4** $1\frac{1}{3}$ m^2

5 a $7\frac{7}{8}$ km b $3\frac{15}{16}$ km c $83\frac{19}{40}$ km

6 $3\frac{1}{40}$ metres **7** $1\frac{1}{6}$ cups **8** $7\frac{1}{6}$ bags

9 a $60\frac{2}{3}$ km b $34\frac{2}{3}$ km c $95\frac{1}{3}$ km d $69\frac{1}{3}$ km

10 $42\frac{1}{6}$ km

11 a $10\frac{33}{40}$ km b $1\frac{23}{40}$ km

12 $4\frac{11}{20}$ kg

13 $9\frac{7}{24}$ cups

Pages 21–22

1 a < b > c < d > e > f <
g > h < i < j > k < l <
m < n < o > p > q < r <
s > t < u < v > w < x >

2 8.2, 8.06, 8.5, 8.09, 8.055, 8.15, 8.051, 8.08, 8.1, 8.053

3 5.17, 5.69, 5.07, 5.65, 5.57, 5.073, 5.08, 5.67, 5.008, 5.1

4 6.25, 6.1, 6.085, 6.17, 6.24, 6.09, 6.119, 6.206, 6.099, 6.211

5 3.07, 2.978, 2.98, 3.114, 3.099, 2.999, 3.105, 3.15, 2.973

6 a 5.06, 5.067, 5.07, 5.17, 5.5, 5.66, 5.67, 5.7
b 2.025, 2.03, 2.138, 2.25, 2.3, 2.38, 2.83, 2.88
c 9.017, 9.09, 9.1, 9.115, 9.189, 9.82, 9.875, 9.9
d 0.04, 0.05, 0.45, 0.455, 0.499, 0.5, 0.54, 0.55
e 4.09, 4.8, 4.85, 4.89, 4.9, 4.98, 4.989, 4.99
f 7.03, 7.032, 7.035, 7.23, 7.3, 7.32, 7.5, 7.55
g 1.06, 1.067, 1.6, 1.67, 1.675, 1.68, 1.7, 1.75
h 3.01, 3.015, 3.1, 3.15, 3.155, 3.25, 3.5, 3.51
i 6.004, 6.04, 6.045, 6.05, 6.4, 6.44, 6.45, 6.5
j 8.026, 8.129, 8.13, 8.16, 8.19, 8.2, 8.25, 8.26
k 2.07, 2.077, 2.6, 2.7, 2.76, 2.77, 2.775, 2.8
l 0.035, 0.3, 0.33, 0.335, 0.34, 0.345, 0.39, 0.4

Pages 22–23

1 a 3.6 b 8.7 c 0.3 d 6.4 e 10.1 f 5.4
g 11.6 h 4.4 i 12.9 j 2.2 k 9.4 l 7

2 a 7.54 b 0.82 c 3.67 d 15.32 e 11.06 f 2.92
g 8.15 h 5.3 i 10.84 j 6.09 k 4.75 l 1.56

3 a 15.715 b 5.864 c 8.076 d 0.456 e 12.225 f 2.318
g 6.581 h 3.499 i 20 j 13.764 k 7.08 l 4.212

4 a 25 b 17 c 9 d 12 e 9 f 4
g 16 h 28 i 37 j 21 k 43 l 20

5 a 4.4, 4, 4.38 b 9.1, 9, 9.11 c 16.8, 17, 16.84
d 10.5, 11, 10.55 e 7.3, 7, 7.29 f 25.6, 26, 25.62
g 0.8, 1, 0.77 h 5.4, 5, 5.44 i 8.1, 8, 8.05
j 13, 13, 12.99 k 30.3, 30, 30.29 l 3, 3, 3.01
m 1.8, 2, 1.77 n 6.3, 6, 6.25 o 11.6, 12, 11.64

Pages 23–24

1 a $5 + 12 = 17$ b $12 - 6 = 6$ c $6 \times 5 = 30$ d $15 \div 3 = 5$
e $15 + 9 = 24$ f $25 - 12 = 13$ g $11 \times 9 = 99$ h $24 - 6 = 18$
i $25 + 17 = 42$ j $34 - 11 = 23$ k $4 \times 8 = 32$ l $36 \div 4 = 9$
m $19 + 17 = 36$ n $41 - 16 = 25$ o $12 \times 8 = 96$ p $27 \div 3 = 9$
q $33 + 25 = 58$ r $30 - 15 = 15$ s $20 \times 5 = 100$ t $40 - 10 = 30$
u $26 + 8 = 34$ v $34 - 8 = 26$ w $13 \times 3 = 39$ x $32 \div 8 = 4$

2 a 63.85 b 58.12 c 55.123 d 100.011 e 150.68 f 49.62
g 20.167 h 2.99 i 361.65 j 127.9 k 46.3 l 90.767
m 74.653 n 97.733 o 6.001 p 8.0 q 131.24 r 576.23
s 301.08 t 29.059 u 184.856 v 988.91 w 5.041 x 8.292
y 343.813 z 203.6

Pages 25–26

1

+	24.8	7.975	47.68	321.7	86.529	6.554	183.65	35.97	277.6
71.263	96.063	79.238	118.943	392.963	157.792	77.817	254.913	107.233	348.863
8.575	33.375	16.55	56.255	330.275	95.104	15.129	192.225	44.545	286.175
164.92	189.72	172.895	212.6	486.62	251.449	171.474	348.57	200.89	442.52
37.96	62.76	45.935	85.64	359.66	124.489	44.514	221.61	73.93	315.56
64.03	88.83	72.005	111.71	385.73	150.559	70.584	247.68	100	341.63
8.074	32.874	16.049	55.754	329.774	94.603	14.628	191.724	44.044	285.674

2 a 533.676 b 377.589 c 431.375 d 487.634 e 500.163 f 615.199

3 a $65.374 + 41.253 = 106.627$ b $4.817 + 27.56 = 32.377$
c $56.08 + 79.864 = 135.944$ d $4.905 + 87.64 = 92.545$
e $39.153 + 58.766 = 97.919$ f $55.96 + 72.85 = 128.81$
g $43.158 + 8.57 = 51.728$ h $9.878 + 97.066 = 106.944$
i $8.674 + 63.886 = 72.560$ j $237.8 + 386.69 = 624.49$
k $67.58 + 18.954 = 86.534$ l $218.07 + 88.635 = 306.705$

4 a 68.15 b 206.86 c 102.035 d 204.797 e 119.87
f 23.96 g 206.82 h 129.104 i 127.43 j 181.59
k 106.239 l 67.976 m 188.96 n 261.22 o 180.817
p 76.483 q 169.24 r 98.098 s 185.38 t 106.166
The place name is KAVIENG.

Pages 27–28

1 a 97.04, 84.19, 59.49, 50.941 b 116.5, 103.65, 78.95, 70.401
c 104.76, 91.91, 67.21, 58.661 d 83.261, 70.411, 45.711, 37.162
e 134.9, 122.05, 97.35, 88.801 f 232.18, 219.33, 194.63, 186.081
g 125.27, 112.42, 87.72, 79.171 h 78.063, 65.213, 40.513, 31.964
i 94.7, 81.85, 57.15, 48.601 j 228.59, 215.74, 191.04, 182.491
k 81.475, 68.625, 43.925, 35.376 l 148.6, 135.75, 111.05, 102.501

2 a 319.19, 265.1, 147.9, 125.528, 81.658, 71.888
b 578.72, 546.82, 468.288, 311.248, 229.588, 182.508
c 517.553, 403.253, 311.043, 280.024, 171.584, 134.334
d 729.25, 671.134, 587.844, 458.974, 384.931, 216.531

3 a 78.18 b 61.22 c 54.66 d 56.78 e 67.72
f 44.735 g 86.23 h 154.41 i 38.236 j 76.62
k 147.25 l 117.857 m 79.32 n 35.878 o 67.86
p 88.23 q 126.293 r 52.592 s 84.835 t 107.62
The place name is RABAUL.

Page 29

1 a 548.47, 5484.7, 54 847 b 25.59, 255.9, 2559
c 391.6, 3916, 39 160 d 789.25, 7892.5, 78 925
e 1250.4, 12 504, 125 040 f 625.8, 6258, 62 580
g 3441.7, 34 417, 344 170 h 639, 6390, 63 900
i 1892, 18 920, 189 200 j 216.25, 2162.5, 21 625
k 30.89, 308.9, 3089 l 2574, 25 740, 257 400
m 159.25, 1592.5, 15 925 n 2420.7, 24 207, 242 070
o 57.03, 570.3, 5703 p 76.8, 768, 7680
q 363.31, 3633.1, 36 331 r 294.76, 2947.6, 29 476

2 a 27 813 m b 587.6 cm c 1875 mm d 5476 cm
e 287.5 mm f 12 472 m g 2468.7 mm h 95 006 m
i 14 035 cm j 84 030 m k 3375 cm l 497.2 mm
m 21 618.2 cm n 1096.1 mm o 257 180 m p 98.16 mm
q 7400 m r 56 544.5 cm s 105 800 m t 1480.7 cm
u 741.9 mm v 5045.2 cm w 1108 mm x 99 510 m

Page 30

1 1807.2 **2** 3459.2 **3** 6566 **4** 6949.2 **5** 6522
6 13 775 **7** 4809.6 **8** 6452.8 **9** 2886.8 **10** 676.41
11 561.12 **12** 293.9 **13** 9372 **14** 3364.2 **15** 9941.6
16 3166.16 **17** 7406.1 **18** 128.38 **19** 11 640 **20** 1932.56
21 19 228.8 **22** 1891.56 **23** 6696 **24** 6741.04 **25** 4072
26 3624 **27** 780.39 **28** 2056.2

Page 31

1 236.4 **2** 80.73 **3** 89.25 **4** 191.1 **5** 324.24
6 1740.8 **7** 65.646 **8** 32.965 **9** 840.6 **10** 726.6
11 164.92 **12** 926.5 **13** 594.75 **14** 2346.5 **15** 194.649
16 110.272 **17** 891.82 **18** 253.663 **19** 4210.6 **20** 677.92
21 858.52 **22** 6256.8

The total of the five largest answers is 15 481.2

Page 32

1 a 184.86 b 234.26 c 265.6964 d 171.69
e 645.12 f 517.4775 g 99.84 h 435.896
i 175.968 j 99.896 k 187.668 l 343.98

2 a 16.718 b 273.22 c 136.812 d 43.0493 e 771.75 f 200.232
g 71.4612 h 32.6534 i 415.492 j 16.4115 k 1190.16 l 343.75
m 423.516 n 530.95 o 716.42

Page 33

1 a 71.86, 7.186, 0.7186 b 967.264, 96.7264, 9.67264
c 43.8427, 4.38427, 0.438427 d 325.78, 32.578, 3.2578
e 31.608, 3.1608, 0.31608 f 20.625, 2.0625, 0.20625
g 153.466, 15.3466, 1.53466 h 41.09, 4.109, 0.4109
i 76.528, 7.6528, 0.76528 j 37.45, 3.745, 0.3745
k 80.954, 8.0954, 0.80954 l 762.94, 76.294, 7.6294
m 2.2093, 0.22093, 0.022093 n 5.413, 0.5413, 0.05413
o 68.82, 6.882, 0.6882 p 1.7551, 0.17551, 0.017551
q 25.3917, 2.53917, 0.253917 r 3.84, 0.384, 0.0384

2 a 45.67 km b 11.65 m c 25.4 cm d 3.827 m
e 6.325 cm f 2.3754 km g 1.449 cm h 0.88512 km
i 6.482 m j 3.8036 km k 4.091 m l 17.85 cm
m 0.5708 m n 7.76 m o 6.328 km p 0.871 cm
q 8.2098 km r 9.68 m s 17.809 km t 0.0574 m
u 913.21 cm v 0.085 m w 150.6 cm x 3.35922 km

Page 34

1 2.817 **2** 11.32 **3** 8.72 **4** 25.73 **5** 12.17
6 4.372 **7** 6.14 **8** 0.356 **9** 15.78 **10** 3.93
11 1.804 **12** 4.57 **13** 32.73 **14** 3.087 **15** 13.29
16 2.155 **17** 41.88 **18** 27.03 **19** 16.47 **20** 28.47

The total of the columns added together does equal the GRAND TOTAL.

Page 35

1 13.15 **2** 133.9 **3** 283.8 **4** 21.58 **5** 14.19
6 146.36 **7** 61.42 **8** 71.16 **9** 37.25 **10** 8.94
11 44.3 **12** 12.08 **13** 47.39 **14** 27.8 **15** 44.7
16 36.14 **17** 43.71 **18** 17.86 **19** 11.85 **20** 45.7
21 31.04 **22** 22.41 **23** 25.04

The lucky ticket is number 14.

Page 36

1 14.63 **2** 21.74 **3** 20.65 **4** 24.5 **5** 18.42
6 13.59 **7** 14.92 **8** 9.94 **9** 67.29 **10** 50.61
11 22.94 **12** 47.28 **13** 70.97 **14** 34.89 **15** 108.09
16 118.63 **17** 56.11 **18** 67.78 **19** 41.45 **20** 42.45

Page 37

Game

Page 38

1 a K153.55 b K1842.60
2 a K33.21 b K20.66 c K39.26 d K27.82
3 K63.85
4 K43.76
5 a 318.2 kg b 53.39 kg
6 K566.54
7 a K815.45 b K556.85
8 a 67.7 km per hour b 609.3 km
9 247.72 km
10 a 469.8 kg b 783 kg c 281.88 kg d 689.04 kg
11 a 3332.92 litres b 833.23 litres

Pages 39–40

1 a 0.54 b 0.7 c 0.31 d 0.325 e 0.503 f 0.63
g 0.9 h 0.437 i 0.04 j 0.067 k 0.194 l 0.014
m 0.8 n 0.04 o 0.829 p 0.05 q 0.13 r 0.346
s 0.007 t 0.45 u 0.087 v 0.009 w 0.01 x 0.001

2 a 0.8 b 0.65 c 0.625 d 0.75 e 0.4 f 0.24
g 0.125 h 0.28 i 0.6 j 0.6 k 0.05 l 0.375
m 0.58 n 0.45 o 0.76 p 0.2 q 0.556 r 0.136
s 0.9 t 0.268 u 0.408 v 0.875 w 0.88 x 0.3

3 a 0.83 b 0.58 c 0.44 d 0.43 e 0.77 f 0.73
g 0.33 h 0.36 i 0.13 j 0.55 k 0.18 l 0.47
m 0.71 n 0.17 o 0.22 p 0.64 q 0.92 r 0.57
s 0.11 t 0.42 u 0.09 v 0.86 w 0.88 x 0.66

4 a 2.3 b 1.53 c 2.007 d 1.85 e 3.9 f 5.39
g 5.08 h 3.244 i 7.61 j 3.095 k 1.6 l 1.375
m 6.875 n 3.48 o 5.45 p 4.75 q 1.875 r 7.4
s 3.05 t 6.68 u 2.84 v 3.375 w 10.75 x 12.1

5 a $\frac{4}{5}$ b $\frac{513}{1000}$ c $\frac{71}{100}$ d $1\frac{1}{2}$ e $3\frac{3}{25}$ f $\frac{243}{1000}$
g $\frac{23}{1000}$ h $5\frac{7}{100}$ i $2\frac{3}{20}$ j $1\frac{9}{1000}$ k $6\frac{1}{5}$ l $8\frac{11}{50}$
m $3\frac{857}{1000}$ n $7\frac{93}{100}$ o $4\frac{3}{5}$ p $5\frac{11}{125}$ q $1\frac{11}{40}$ r $\frac{31}{200}$
s $2\frac{13}{20}$ t $4\frac{81}{250}$ u $9\frac{2}{125}$ v $1\frac{17}{20}$ w $3\frac{3}{500}$ x $2\frac{209}{500}$

Pages 40–42

1 a 27% b 63% c 11% d 53% e 68% f 41%
g 155% h 7% i 208% j 165% k 233% l 83%
m 219% n 105% o 40% p 32% q 16% r 70%
s 180% t 224% u 135% v 70% w 140% x 62%

2 a 90% b 131% c 25% d 80% e 15% f 208%
g 175% h 150% i 673% j 284% k 0.7% l 120%
m 94% n 30% o 56% p 62.5% q 40% r 78%
s 161% t 28% u 503.5% v 460% w 170% x 1%

3 a 0.57 b 0.32 c 0.73 d 0.14 e 0.07 f 0.94
g 0.81 h 1.63 i 0.04 j 2.17 k 0.28 l 1.35
m 0.545 n 0.4 o 0.245 p 0.1725 q 0.3975 r 0.86
s 0.01 t 1 u 2.09 v 1.46 w 0.9 x 0.3325

4 a $\frac{1}{10}$ b $\frac{2}{5}$ c $\frac{3}{50}$ d $\frac{7}{10}$ e $\frac{3}{20}$ f $\frac{12}{25}$
g $\frac{16}{25}$ h $\frac{13}{50}$ i $\frac{18}{25}$ j $\frac{19}{50}$ k $\frac{19}{20}$ l $1\frac{1}{5}$
m $2\frac{1}{2}$ n $1\frac{3}{4}$ o $\frac{7}{50}$ p $\frac{33}{50}$ q $\frac{1}{25}$ r $\frac{43}{50}$
s $\frac{67}{200}$ t $\frac{7}{40}$ u $\frac{91}{200}$ v $\frac{5}{8}$ w $\frac{9}{80}$ x $\frac{29}{80}$

5 a

Fraction	Decimal	Percentage
$\frac{4}{5}$	0.8	80%
$\frac{2}{5}$	0.4	40%
$\frac{9}{100}$	0.09	9%
$\frac{1}{4}$	0.25	25%
$\frac{7}{8}$	0.875	$87\frac{1}{2}$%
$\frac{17}{20}$	0.85	85%
$\frac{9}{20}$	0.45	45%
$\frac{1}{8}$	0.125	$12\frac{1}{2}$%
$\frac{7}{25}$	0.28	28%
$\frac{7}{10}$	0.7	70%
$1\frac{7}{100}$	1.07	107%
$\frac{1}{200}$	0.005	0.5%

b

Fraction	Decimal	Percentage
$\frac{3}{4}$	0.75	75%
$\frac{14}{25}$	0.56	56%
$\frac{3}{8}$	0.375	$37\frac{1}{2}$%
$\frac{3}{100}$	0.03	3%
$\frac{15}{50}$	0.3	30%
$\frac{1}{100}$	0.01	1%
$1\frac{2}{5}$	1.4	140%
$1\frac{3}{5}$	1.6	160%
$1\frac{71}{100}$	1.71	171%
$\frac{1}{2}$	0.5	50%
$\frac{1}{125}$	0.008	0.8%
$2\frac{3}{8}$	2.375	$237\frac{1}{2}$%

6 a $\frac{6}{10}$ b $33\frac{1}{3}$% c 75% d $\frac{27}{100}$ e $\frac{1}{2}$ f 0.7
g 25% h $\frac{20}{100}$ i $\frac{3}{4}$ j $\frac{1}{10}$ k 0.2 l 30%
m $66\frac{2}{3}$% n 0.625 o 0.9%

7 a 0.09 b 0.2 c 50% d 70% e 6% f $\frac{2}{4}$
g 0.6 h 0.85 i 0.14 j 6% k $\frac{7}{10}$ l 1%
m $3\frac{1}{2}$% n 42.5 o 17%

Page 43

1 a K9, K81 b K11, K44 c K1.25, K23.75
d K45, K105 e K12, K36 f K12.50, K87.50
g K4, K76 h K7.50, K92.50 i K43.75, K131.25
j K18, K102 k K3.75, K33.75 l K12.50, K50
m K30, K210 n K14.40, K81.60 o K8.25, K156.75
p K10.50, K129.50 q K2.35, K21.15 r K21.25, K63.75
s K16.50, K93.50

2 a K75, K225 b K39, K221 c K8.75, K166.25
d K21.50, K193.50 e K22.80, K262.20 f K4.40, K105.60
g K57.75, K272.25 h K40.50, K229.50 i K81, K324
j K48.75, K341.25 k K13.20, K206.80 l K8.40, K103.60
m K51.75, K293.25 n K108.50, K201.50 o K68.75, K206.25
p K28, K132 q K47.20, K188.80 r K18.75, K168.75
s K37.35, K87.15

Pages 44–45

1 a K7.50, K57.50 b K21.25, K106.25 c K6.30, K69.30
d K18, K162 e K165, K715 f K75, K450
g K234, K1794 h K142.50, K2992.50 i K134.50, K1210.50
j K582.50, K2912.50

2 a 207 g b 172.5 g c 241.5 g d 575 g
e 414 g f 483 g g 839.5 g h 1265 g
i 2990 g j 2242.5 g k 3737.5 g l 2587.5 g
m 2150.5 g n 2357.5 g o 2058.5 g p 3542 g
q 3392.5 g r 1794 g

3 a 1008 mL b 738 mL c 1212 mL d 870 mL
e 1500 mL f 810 mL g 2808 mL h 2202 mL
i 3588 mL j 3750 mL k 3324 mL l 966 mL
m 1194 mL n 2412 mL o 4410 mL p 4806 mL
q 2742 mL r 1098 mL

4 a 157.5 cm b 357 cm c 283.5 cm d 808.5 cm
e 588 cm f 430.5 cm g 934.5 cm h 682.5 cm
i 976.5 cm j 1302 cm k 1690.5 cm l 2268 cm
m 1134 cm n 3412.5 cm o 2446.5 cm p 1837.5 cm
q 4084.5 cm r 2142 cm

5 a K60.20 b K133.30 c K98.90 d K266.60
e K202.10 f K215 g K365.50 h K118.25
i K264.45 j K397.75 k K541.80 l K453.65
m K1075 n K2365 o K1806 p K3182
q K1494.25 r K4095.75

Pages 45–46

1 K113.05

2 a 92 guava b 115 mango c 253 orange

3 3870 adults

4 a K232.50 bonus b K4882.50 in total

5 104 people

6 a 250 g flour b 175 g sugar c 87.5 g cocoa d 150 g butter

7 a K103 200 b K90 300 c K98 900 d K96 750
e K92 020 f K87 720

8 207 students

9 a 9375 litres b 3125 litres

10 K16 032

11 147 cm

12 K7.35

13 a K9072 b K24 528

14 K778.75

15 K714.45

16 30 625 spectators

17 14 litres

18 K35 475

19 142.5 minutes

20 a 51 students b 289 students

21 162.75 cm

22 K45 000

23 a K4368 b K3974.88

24 a 143 girls b 117 boys

Pages 47–48

1 a 2:5, $\frac{2}{5}$ b 5:7, $\frac{5}{7}$ c 7:2, $\frac{7}{2}$ d 2:14, $\frac{2}{14}$
e 7:14, $\frac{7}{14}$ f 5:14, $\frac{5}{14}$

2 a 15:6, $\frac{15}{6}$ b 7:15, $\frac{7}{15}$ c 15:28, $\frac{15}{28}$ d 6:7, $\frac{6}{7}$
e 7:28, $\frac{7}{28}$ f 6:28, $\frac{6}{28}$

3 a 9:5, $\frac{9}{5}$ b 8:11, $\frac{8}{11}$ c 11:5, $\frac{11}{5}$ d 5:8, $\frac{5}{8}$
e 11:33, $\frac{11}{33}$ f 8:33, $\frac{8}{33}$

4 a 1:7, $\frac{1}{7}$ b 10:3, $\frac{10}{3}$ c 8:17, $\frac{8}{17}$ d 5:2, $\frac{5}{2}$
e 5:50, $\frac{5}{50}$ f 12:27, $\frac{12}{27}$ g 30:85, $\frac{30}{85}$ h 45:8, $\frac{45}{8}$
i 8:20, $\frac{8}{20}$ j 12:5, $\frac{12}{5}$ k 25:20, $\frac{25}{20}$ l 15:18, $\frac{15}{18}$
m 24:16, $\frac{24}{16}$ n 3:16, $\frac{3}{16}$ o 14:20, $\frac{14}{20}$ p 30:45, $\frac{30}{45}$

5 a 1:3 b 2:7 c 5:3 d 7:10 e 4:1 f 5:0
g 3:7 h 9:8 i 10:3 j 5:9 k 11:6 l 1:8
m 1:2 n 4:3 o 12:7 p 8:11 q 9:15 r 6:1

6 a $\frac{6}{11}$ b $\frac{4}{5}$ c $\frac{7}{4}$ d $\frac{13}{15}$ e $\frac{5}{7}$ f $\frac{8}{5}$
g $\frac{3}{9}$ h $\frac{12}{5}$ i $\frac{1}{5}$ j $\frac{10}{9}$ k $\frac{8}{17}$ l $\frac{11}{4}$
m $\frac{5}{8}$ n $\frac{9}{20}$ o $\frac{6}{7}$ p $\frac{5}{12}$ q $\frac{19}{8}$ r $\frac{15}{21}$

Pages 48–49

1 a 2:3 b 1:4 c 3:2 d 1:3 e 3:2 f 1:3
g 1:5 h 1:5 i 2:1 j 1:5 k 1:2 l 3:1
m 1:3 n 6:5 o 7:8 p 1:5 q 2:5 r 4:3
s 1:4 t 3:1 u 3:5 v 2:3 w 5:4 x 2:3

2 a 1:4 b 2:1 c 1:6 d 3:4 e 1:3 f 3:8
g 11:2 h 15:6 i 3:2 j 2:9 k 8:3 l 5:7
m 7:12 n 5:2 o 3:4 p 4:3 q 3:10 r 5:3

3 a 3:1 b 1:2 c 3:5 d 7:5 e 3:1 f 2:1
g 2:3 h 1:2

4 a 1:3 b 3:5 c 1:5 d 5:12 e 2:3 f 2:3
g 1:3 h 2:5

Page 50

1 a 1:10 b 1:1000 c 10:1 d 1:5 e 10:1 f 5:1
g 5:9 h 5:2 i 1:2 j 1:12 k 4:3 l 1:8
m 5:2 n 1:20 o 1:15 p 1:6 q 6:1 r 1:10
s 2:5 t 13:20 u 4:1

2 a 1:100 000 b 1:500 c 1:250
d 1:20 000 e 1:500 f 1:50 000
g 1:40 000 h 1:1000 i 1:10 000
j 1:1000 k 1:10 000 l 1:12 500
m 1:50 000 n 1:100 o 1:2500
p 1:20 000

3 a 30:7 b 7:30

4 a 15:4 b 4:15

5 a 84 b 45

6 a 30 cm b 32 cm c 27 cm d 24 cm

Page 51

1 a > b < c < d > e < f >
g < h < i < j < k < l >
m > n < o < p < q > r >
s > t <

2 a < b < c < d < e < f >
g > h < i < j > k > l >
m < n > o > p > q > r >
s > t >

3 a –73 b –58 c –81 d –76 e –96 f –39
g –14 h –43 i –35 j –89 k –32 l –17

4 a +12, +3, –1, –8, –15 b +31, +17, 0, –5, –22
c 27, 23, +14, –9, –18 d 38, +16, –2, –7, –13
e 35, +1, –3, –21, –43 f +48, 42, 36, 53, –59
g 32, +29, 25, –26, –29 h +6, 0, –5, –6, –12
i 55, +44, 43, –40, –51 j 95, +91, +88, –67, –72
k 72, +63, 59, –60, –65 l 47, 38, –36, –41, –48
m +8, 0, –5, –10, –14 n 87, +81, 77, –73, –82
o +56, 49, +28, –42, –45

Page 52

1 a –5 b –1 c +2 d –3

2 a +4 b –3 c –6 d –7 e +2 f –4
g –3 h –3

3 7°

4 a 7°C b 3°C c –3°C d –1°C e 2°C f –7°C

5 Amounts owing: January = K131, February = K104, March = K149, April = K124, May = K88, June = K58, July = K102, August = K70, September = K123, October = K105, November = K69, December = K45

Pages 53–54

1 a 7^6 b 4^5 c 2^8 d 5^7 e 8^5 f 3^9
g 6^4 h 9^3 i 4^7 j 2^5 k 3^8 l 6^{10}
m 10^4 n 8^7 o 5^6 p 11^5

2 a 5 × 5 × 5 b 7 × 7 × 7 × 7
c 2 × 2 d 9 × 9 × 9 × 9 × 9 × 9 × 9
e 3 × 3 × 3 × 3 × 3 × 3 × 3 × 3 × 3 f 4 × 4 × 4 × 4 × 4 × 4
g 2 × 2 × 2 × 2 × 2 h 8 × 8 × 8
i 6 × 6 × 6 × 6 j 5 × 5 × 5 × 5 × 5 × 5 × 5
k 9 × 9 l 10 × 10 × 10 × 10
m 1 × 1 × 1 × 1 × 1 × 1 × 1 × 1 × 1 n 3 × 3 × 3 × 3 × 3
o 7 × 7 × 7 × 7 × 7 × 7 × 7 × 7 p 11 × 11 × 11 × 11
q 6 × 6 × 6 × 6 × 6 × 6 × 6 × 6 × 6 r 4 × 4 × 4 × 4 × 4 × 4 × 4 × 4

3 a 59 b 744 c 150 d 32 668 e 2025 f 2782
g 1574 h 80 i 513 j 135 k 2450 l 1215
m 868 n 900 o 6686 p 169

4 a 1000 b 432 c 4000 d 5400
e 25 600 f 16 384 g 11 664 h 1 605 632
i 35 937 j 164 025 k 124 416 l 512 000
m 110 592 n 605 052 o 419 904 p 28 125

5 a < b > c < d > e > f >
g < h < i < j < k < l >
m > n < o < p >

6 a $3^3 = 27$ b $5^3 = 125$ c $7^3 = 343$ d $2^8 = 256$
e $6^4 = 1296$ f $7^5 = 16\,807$ g $2^9 = 512$ h $9^3 = 729$
i $8^3 = 512$ j $11^4 = 14\,641$ k $7^6 = 117\,649$ l $8^8 = 16\,777\,216$
m $3^9 = 19\,683$ n $4^5 = 1024$ o $4^6 = 4096$ p $3^8 = 6561$

7 a 6^7 b 4^2 c $5^4, 4^2, 3^5, 9^3, 2^7$
d $6^7, 10^4, 6^5, 3^8$ e 2^7 f $5^4, 9^3, 12^3$
g 9^3 and 2^7 h $6^7 - 10^4 = 269\,936$

8 a 5^3 b 6^6 c $6^6, 9^4, 3^7, 11^3, 8^5$
d 10^3 e $3^6, 11^3, 2^9, 10^3$ f $4^4, 5^3$
g 3^7 and 11^3 h $6^6 - 8^5 = 13\,888$

9 a 189 b 2793 c 323 d 2658 e 1152 f 553
g 914 h 2876 i 3274 j 601 k 34 532 l 65 649

10 a < b < c < d > e < f <
g < h < i > j > k > l >

Pages 55–57 Assessment

1 a $\frac{2}{6}$ b $\frac{12}{28}$ c $\frac{2}{9}$ d $\frac{28}{48}$

2 a $\frac{3}{4}$ b $\frac{4}{9}$ c $\frac{4}{5}$ d $\frac{1}{7}$ e $\frac{3}{10}$ f $\frac{5}{11}$

3 a $\frac{21}{5}$ b $\frac{21}{8}$ c $\frac{23}{4}$ d $\frac{5}{3}$ e $\frac{23}{6}$ f $\frac{20}{3}$

4 a $5\frac{1}{3}$ b $4\frac{5}{6}$ c $2\frac{3}{7}$. d $3\frac{4}{9}$ e $4\frac{2}{5}$ f $4\frac{5}{8}$

5 a $1\frac{1}{6}$ b $\frac{1}{3}$ c $4\frac{1}{2}$ d $1\frac{1}{2}$ e $5\frac{21}{40}$ f $1\frac{16}{21}$

6 a $\frac{4}{21}$ b $2\frac{5}{8}$ c $24\frac{3}{4}$ d $\frac{17}{30}$ e 4 f $2\frac{19}{22}$

7 a $109\frac{23}{24}$ km b $2\frac{23}{24}$ km c $40\frac{31}{40}$ kg

8 a < b < c > d > e > f <

9 a 4.275, 4.21, 4.2, 4.16, 4.07, 4.033, 4.025, 4.02
b 7.9, 7.86, 7.816, 7.8, 7.19, 7.09, 7.082, 7.08
c 2.56, 2.5, 2.45, 2.425, 2.4, 2.05, 2.005, 2.004

10

Decimal number	Rounded to one decimal place	Rounded to two decimal places	Rounded to nearest whole number
4.116	4.1	4.12	4
8.552	8.6	8.55	9
5.073	5.1	5.07	5
2.855	2.9	2.86	3
6.007	6	6.01	6

11 a 42.71 b 84.223 c 225.45 d 186.109 e 45.18 f 77.81
g 113.43 h 57.44

12 a 3640 b 572.3 c 262.71 d 4887 e 527.1 f 140.76
g 341.744 h 1373.42

13 a 7.612 b 15.386 c 40.84 d 2.747 e 134.9 f 48.14
g 27.9 h 63.6

14 a 5:3, $\frac{5}{3}$ b 2:5, $\frac{2}{5}$ c 3:10, $\frac{3}{10}$

15 a 8.77 kg b 31.635 kg c K612

16 a 2:5 b 9:7 c 1:4 d 6:1 e 2:9 f 10:3

17 a 1:5 b 1:4 c 20:1 d 10:1 e 3:10 f 1:3

18 a < b > c > d > e < f <
g > h <

19 a −14, −8, −6, +10, +17 b −16, −11, 0, 5, +9
c −17, −12, +6, +17, 20 d −62, −50, −23, 57, +63
e −41, −35, −29, +37, 38 f −92, −87, −86, +71, 80

20 a −K7 b −K7 c K3 d K55

21 a 9^7 b 12^5 c 25^3 d 16^6

22 a 43 b 89 c 2048 d 972 e 1299 f 27 648
g 2251 h 2224

23 a > b > c >

Page 58 Assessment

1 b **2** c **3** b **4** a **5** d **6** b
7 d **8** b **9** a **10** c **11** b **12** c
13 d **14** d **15** a **16** b **17** c **18** d
19 a

Strand: Space and Shape

Page 59

1 6000 mm, 250 cm, 68 cm, 55 cm, 510 mm, 550 mm, 60 cm, 506 mm

2 150 cm, 1600 mm, 15 cm, 20 cm, 500 mm, 197 cm, 1950 mm, 1995 mm

3 a 38 mm b 15 cm c 125 cm d 9.5 mm e 0.58 m f 4.25 m
g 0.89 m h 410 mm i 215 cm j 1.5 cm k 0.7 km l 0.6 km

4 a 0.4 km, 17 m, 32 mm, 2.5 cm b 0.52 km, 50 m, 50 cm, 50 mm
c 65 km, 6500 cm, 6.5 m, 0.65 m d 0.76 km, 8700 cm, 76 m, 87 cm
e 0.35 km, 350 mm, 3.55 cm, 0.035 m f 9300 m, 0.93 km, 9300 cm, 9.3 m

5 a 16 300 m or 16.3 km b 279.3 cm c 4.675 km
d 87 cm e 2.15 km

Pages 60–61

1 9.42 cm **2** 15.7 cm **3** 12.56 cm **4** 37.68 cm **5** 25.12 cm
6 21.98 cm **7** 28.26 cm **8** 20.41 cm **9** 32.97 cm **10** 29.83 cm
11 14.13 cm

Page 62

1 a 62.8 cm b 47.1 cm c 31.4 cm d 50.24 cm
e 72.22 cm f 34.54 cm g 56.52 cm h 78.5 cm
i 94.2 cm j 157 cm k 54.95 cm l 29.05 cm
m 24.34 cm n 39.25 cm o 41.49 cm p 68.14 cm
q 60.6 cm r 27.95 cm

2 a 31.4 cm b 81.64 cm c 25.12 cm d 106.76 cm
e 50.24 cm f 87.92 cm g 100.48 cm h 59.66 cm
i 78.5 cm j 47.1 cm k 36.11 cm l 72.85 cm
m 51.81 cm n 93.57 cm o 42.7 cm p 105.19 cm
q 65 cm r 24.18 cm

3 a 7 cm, 14 cm, 43.96 cm b 11.5 cm, 23 cm, 72.22 cm
c 6.35 cm, 12.7 cm, 39.88 cm d 7.7 cm, 15.4 cm, 48.36 cm
e 4.1 cm, 8.2 cm, 25.75 cm f 3.15 cm, 6.3 cm, 19.782 cm
g 11.7 cm, 23.4 cm, 73.48 cm h 13.65 cm, 27.3 cm, 85.72 cm
i 9.18 cm, 18.36 cm, 57.65 cm j 4.8 cm, 9.6 cm, 30.144 cm
k 16.75 cm, 33.5 cm, 105.19 cm l 13.25 cm, 26.5 cm, 83.21 cm
m 4.85 cm, 9.7 cm, 30.46 cm n 6.35 cm, 12.7 cm, 39.88 cm
o 5.25 cm, 10.5 cm, 32.97 cm p 10.3 cm, 20.6 cm, 64.684 cm
q 11.55 cm, 23.1 cm, 72.534 cm r 14.65 cm, 29.3 cm, 92 cm
s 8.45 cm, 16.9 cm, 53.07 cm

Page 63

1 a 20 m^2 b 24.5 m^2 c 65 m^2 d 28.75 m^2
e 22.5 m^2 f 90.25 m^2 g 78.75 m^2 h 52.5 m^2

2 a 10 cm^2 b 10.5 cm^2 c 8.75 cm^2 d 7.5 cm^2
e 7 cm^2 f 15.75 cm^2 g 18 cm^2

Page 64

1 17.5 cm^2 **2** 7.5 cm^2 **3** 27 cm^2 **4** 16 cm^2 **5** 44 cm^2
6 54 cm^2 **7** 31.5 cm^2 **8** 42.5 cm^2 **9** 14.25 cm^2 **10** 45 cm^2
11 13.75 cm^2 **12** 47.25 cm^2 **13** 33.75 cm^2 **14** 60.5 cm^2

Pages 65–66

1 18 cm^2 **2** 24 cm^2 **3** 30 cm^2 **4** 104 cm^2 **5** 85.5 cm^2
6 250 cm^2 **7** 24.75 cm^2 **8** 415 cm^2 **9** 68 cm^2 **10** 228.75 cm^2
11 238.5 cm^2 **12** 325 cm^2 **13** 61.75 cm^2 **14** 1050 cm^2 **15** 323.75 cm^2
16 1360 cm^2 **17** 447 cm^2 **18** 1083 cm^2

Pages 67–68

1 48 cm^2 **2** 40 cm^2 **3** 96 cm^2 **4** 65 cm^2
5 42 cm^2 **6** 90 cm^2 **7** 562.5 cm^2 **8** 142.5 cm^2
9 192 cm^2 **10** 800 cm^2 **11** 88 cm^2 **12** 148.5 cm^2
13 345 cm^2 **14** 338 cm^2 **15** 330 cm^2 **16** 126 cm^2
17 297 cm^2 **18** 797.5 cm^2 **19** 331.25 cm^2 **20** 367.5 cm^2
21 625 cm^2 **22** 259.25 cm^2 **23** 1186.25 cm^2

Pages 69–70

1 12 cm^2 **2** 10 cm^2 **3** 39 cm^2 **4** 174 cm^2 **5** 54 cm^2
6 60 cm^2 **7** 32.5 cm^2 **8** 198 cm^2 **9** 172.5 cm^2 **10** 300 cm^2
11 336 cm^2 **12** 48 cm^2 **13** 68 cm^2 **14** 144 cm^2 **15** 140 cm^2
16 40 cm^2 **17** 126.5 cm^2 **18** 240 cm^2 **19** 222 cm^2 **20** 162 cm^2
21 288 cm^2 **22** 348 cm^2

Pages 71–72

1 100 m^2 **2** 96 m^2 **3** 117 m^2 **4** 71 m^2 **5** 86 m^2
6 76 m^2 **7** 73.5 m^2 **8** 95 m^2 **9** 150 m^2 **10** 138 m^2
11 57.5 m^2 **12** 106 m^2 **13** 328 m^2 **14** 240 m^2 **15** 28 m^2
16 147 m^2 **17** 184.5 m^2 **18** 76.5 m^2 **19** 137.5 m^2 **20** 222.5 m^2
21 104 m^2

Pages 73–74

1 a 24 cm^3 b 27 cm^3 c 12 cm^3 d 24 cm^3 e 45 cm^3 f 36 cm^3
g 64 cm^3 h 135 cm^3 i 210 cm^3 j 70 cm^3 k 198 cm^3

2 a 30 cm^3 b 10 cm c 4 cm d 7 cm e 360 cm^3 f 4 cm
g 4 cm h 3 cm i 12 cm j 2 cm k 2 cm l 3 cm
m 10 cm n 432 cm^3 o 5 cm p 288 cm^3 q 7 cm r 6 cm
s 8 cm t 330 cm^3

3 a 65 cm^3 b 90 cm^3 c 21 cm^3 d 4 cm
e 11 cm f 8 cm g 4.5 cm h 202.5 cm^3
i 12 cm j 5 cm k 8 cm l 208.8 cm^3
m 364.5 cm^3 n 10 cm o 8 cm p 2 cm
q 312 cm^3 r 12 cm s 4 cm t 560 cm^3

4 a 150 cm^3 b 216 m^3 c 160 m^3 d 120 cm^3
e 729 cm^3 f 440 cm^3 g 264 cm^3 h 147 m^3
i 693 cm^3 j 1728 cm^3 k 360 m^3 l 48 cm^3
m 882 cm^3 n 1200 m^3 o 125 m^3 p 240 cm^3
q 768 m^3 r 315 cm^3 s 20 cm^3 t 343 cm^3
u 143.75 cm^3 v 535.5 m^3 w 418.5 cm^3 x 1197 m^3

Page 75

1 112 cm^3 **2** 210 cm^3 **3** 12 cm^3 **4** 120 cm^3 **5** 123.75 cm^3
6 180 cm^3 **7** 127.5 cm^3 **8** 126 cm^3 **9** 75 cm^3 **10** 243 cm^3
11 103.5 cm^3 **12** 400 cm^3

Page 76

1 252 cm^3 **2** 315 cm^3 **3** 252 cm^3 **4** 500 cm^3 **5** 462 cm^3
6 429 cm^3 **7** 884 cm^3 **8** 562.5 cm^3 **9** 756 cm^3 **10** 1210 cm^3
11 480 cm^3 **12** 627 cm^3

Pages 77–78

1 2160 cm^3 **2** 2250 cm^3 **3** 3504 cm^3 **4** 3456 cm^3 **5** 997.5 m^3
6 1375 m^3 **7** 1056 m^3 **8** 540 m^3 **9** 1525 m^3 **10** 1320 m^3
11 2322 m^3 **12** 2591 cm^3 **13** 2050 cm^3 **14** 592 cm^3 **15** 840 cm^3
16 455 cm^3

Page 79

1 a 10 cm^3 b 250 cm^3 c 3000 cm^3 d 700 cm^3
e 5000 cm^3 f 1500 cm^3 g 475 cm^3 h 2250 cm^3
i 85 cm^3 j 830 cm^3 k 1750 cm^3 l 9500 cm^3
m 3400 cm^3 n 12 000 cm^3 o 374 cm^3 p 15 000 cm^3

2 a 20 mL b 65 mL c 1.2 L d 250 mL e 4 L f 1.6 L
g 715 mL h 535 mL i 7.5 L j 325 mL k 20 L l 17 L
m 116 mL n 8.3 L o 14.5 L

3 a 230 cm^3 b 180 cm^3 c 6000 cm^3 d 95 cm^3
e 710 cm^3 f 9000 cm^3 g 5500 cm^3 h 1360 cm^3
i 538 cm^3 j 2750 cm^3 k 4100 cm^3 l 755 cm^3
m 6700 cm^3 n 107 cm^3 o 1146 cm^3

4 a 650 mL b 700 mL c 1 L d 3 L e 285 mL f 3.5 L
g 410 mL h 45 mL i 90 mL j 935 mL k 7.6 L l 1.25 L
m 4.4 L n 11 L o 13.5 L

5 a 2 m^3 b 5 m^3 c 1.5 m^3 d 8.5 m^3 e 3.5 m^3 f 6.25 m^3
g 2.75 m^3 h 10 m^3 i 15 m^3 j 27 m^3 k 18.5 m^3 l 16.5 m^3
m 12.25 m^3 n 33.25 m^3 o 13.75 m^3

6 a 3000 L b 8000 L c 5000 L d 2500 L e 4500 L f 10 000 L
g 11 500 L h 14 000 L i 7250 L j 10 250 L k 9750 L l 15 750 L
m 20 000 L n 35 000 L o 26 000 L

Page 80

1 1 L **2** 288 mL **3** 600 mL **4** 222 mL **5** 180 mL
6 247.5 mL **7** 480 mL **8** 292.5 mL **9** 3.696 L **10** 440 mL
11 500 mL **12** 237.5 mL

Page 81

1 A, C, D, F, H, K and L are parallelograms. **2** C, F and L are rhombuses.
3 B, G and J are trapeziums. **4** C, E, F, I and L are kites.

Page 82

1 A = rectangle, 2 axes of symmetry B = square, 4 axes of symmetry
C = parallelogram, no axes of symmetry D = parallelogram, no axes of symmetry
E = rectangle, 2 axes of symmetry F = parallelogram, no axes of symmetry
G = rhombus, 2 axes of symmetry H = square, 4 axes of symmetry
I = rhombus, 2 axes of symmetry J = rectangle, 2 axes of symmetry

2 2 axes of symmetry **3** 4 axes of symmetry
4 2 axes of symmetry **5** No axes of symmetry

Page 83

B, D, E, F, H, I, K and L are parallelograms.

Page 84

1 40° **2** 60° **3** 35° **4** 70° **5** 55° **6** 55°
7 35° **8** 25° **49** 25° **10** 105° **11** 25° **12** 60°

Page 85

1 95° **2** 120° **3** 125° **4** 110° **5** 100° **6** 85°
7 75° **8** 100° **9** 145° **10** 80° **11** 75° **12** 45°

Page 86

1 140° **2** 135° **3** 125° **4** 75° **5** 55° **6** 120°
7 120° **8** 105° **9** 140° **10** 125° **11** 100° **12** 125°

Page 87

1 140° **2** 120° **3** 125° **4** 100° **5** 60° **6** 125°
7 115° **8** 125° **9** 95° **10** 90° **11** 135°

Page 88

1 a acute b obtuse c acute d acute e obtuse f acute
g obtuse h right i obtuse j acute k acute l obtuse
m obtuse n acute o acute

2 a – o Estimates will vary.

3 a 40° b 120° c 80° d 60° e 100° f 70°
g 130° h 90° i 140° j 20° k 50° l 110°
m 160° n 65° o 10°

Page 89

1 20° **2** 60° **3** 50° **4** 80° **5** 70° **6** 45°
7 55° **8** 45° **9** 35° **10** 40° **11** 25° **12** 45°

Page 90

1 a 35° b 12° c 51° d 69° e 76° f 28°
g 5° h 64° i 83° j 48° k 17° l 31°

2 a 70° b 45° c 35° d 52° e 63° f 17°
g 28° h 47° i 33° j 41° k 54° l 13°

Page 91

1 a 90° b 105° c 66° d 45° e 70° f 117°
g 53° h 34° i 93° j 78° k 29° l 121°

2 a 75° b 120° c 145° d 103° e 99° f 27°
g 116° h 68° i 139° j 112° k 43° l 164°

Page 92

1 95° **2** 150° **3** 100° **4** 300° **5** 85° **6** 230°
7 60° **8** 115° **9** 130° **10** 340° **11** 115° **12** 75°
13 95° **14** 55° **15** 10°

Page 93

1 70° **2** 70° **3** 50° **4** 27° **5** 24° **6** 79°
7 320° **8** 103° **9** 32° **10** 34° **11** 48° **12** 234°
13 38° **14** 105° **15** 160°

Page 94

1 070° **2** 240° **3** 105° **4** 310° **5** 150° **6** 330°
7 025° **8** 200° **9** 135°

Page 95

1 a A = N30°E, B = S50°E, C = S35°W, D = N35°W
b A = N75°E, B = S10°E, C = S65°W, D = N40°W
c A = N55°E, B = S35°E, C = S60°W, D = N70°W
d A = N15°E, B = S30°E, C = S70°W, D = N45°W
e A = N50°E, B = S80°E, C = S20°W, D = N5°W
f A = N68°E, B = S54°E, C = S51°W, D = N28°W

2 a S55°E b N35°E c S30°E d S65°W e N80°E f S35°W
g N30°W h S10°W i N85°W j N56°E k S8°E l N50°W
m S76°E n N62°W o N21°W p N27°E

Page 96

1 a 1:500 000 b 1:50 000 c 1:300 000 d 1:1000
e 1:600 f 1:250 g 1:50 000 h 1:25 000
i 1:250 j 1:200 000 k 1:40 000 l 1:2000

2 a 100 m b 200 m c 350 m d 500 m e 25 m f 175 m
g 425 m h 75 m i 600 m j 775 m k 310 m l 580 m

3 a 6 km b 16 km c 20 km d 12 km e 1 km f 50 km g 32 km h 60 km i 15 km j 25 km k 22.5 km l 49.5 km

4 a 100 km b 175 km c 300 km d 125 km e 37.5 km f 375 km g 87.5 km h 262.5 km i 155 km j 70 km k 285 km l 240 km

5 a 2 cm b 5 cm c 0.5 cm d 1.5 cm e 10 cm f 50 cm g 8.5 cm h 12.5 cm i 25 cm j 8 cm k 2.3 cm l 6.1 cm

6 a 1 cm b 4 cm c 8 cm d 7 cm e 10 cm f 0.5 cm g 20 cm h 15 cm i 25 cm j 17 cm k 13 cm l 22 cm

7 a 2 cm b 0.5 cm c 2.5 cm d 4 cm e 10 cm f 30 cm g 8 cm h 10.5 cm i 12.5 cm j 25 cm k 6.5 cm l 11 cm

Page 97

1 a (2, 3) b (4, 1) c (2, –2) d (–3, –2) e (4, –1) f (–3, 2) g (–1, –3) h (–2, 1) i (4, –2) j (–1, 3) k (3, 4) l (2, –4)

2 a A b D c I d F e W f G g O h C i J j U k Z l Y

3 a RABAUL b KIMBE c MADANG d DARU e WEWAK f MENDI g KEREMA h POMIO i ALOTAU j TUFI k TELEFOMIN l POPONDETTA m KUNDIAWA n NAMATANAI o AMBUNTI

4 a (2, 1) (–4, –3) (4, 1) (–2, 1) (–3, 2) (2, 1) (–1, 1) (–4, 2) (2, 1)
b (–1, 3) (–1, –3) (–2, –4) (1, 2) (–1, –3) (2, 1) (–1, 1) (2, 1) (–1, 3) (3, –3)
c (–4, 2) (–1, 3) (3, –3) (3, –1) (–1, 3) (–1, –3) (4, 1) (–4, 2) (2, 1)
d (–2, 1) (4, –2) (2, 1) (–4, 2) (–1, 1) (2, 1) (–1, 3) (3, –3)
e (2, 3) (4, –2) (–4, 2) (–1, 1) (–4, 2) (2, 3) (2, 3) (–4, 2) (–1, 3) (–1, –3) (4, 1)
f (–2, 4) (2, 1) (2, 3) (2, 1) (–1, 3)
g (–4, 3) (4, –2) (–4, 2) (–1, 3) (2, 1)
h (–4, –1) (–4, 2) (–1, –3) (–2, 1) (–1, 3) (2, 1) (2, –2)
i (2, 3) (2, 1) (4, –1) (–4, 2) (4, 1) (–2, 1) (2, 1) (–1, 3)
j (2, –2) (2, 1) (–1, 1) (2, 1) (1, –3) (4, 1) (–4, 2) (2, 1)
k (2, –2) (3, –1) (–3, 2) (3, –1) (–4, 3) (–4, 3) (3, –1)
l (–1, –3) (–2, 1) (4, –2) (–4, 2) (3, –1) (2, 3) (–4, 2) (2, 1)
m (2, 1) (–3, 2) (3, 2) (–1, –3) (–1, 3) (–2, 1) (–4, 2) (–1, 3) (2, 1)
n (–4, –1) (–1, –3) (–1, 3) (–1, –3) (1, 2) (–4, –3) (–1, –3) (–1, 1) (2, 1)
o (–3, –2) (–3, 2) (2, 1) (1, 2) (–4, 2) (–1, 1)

Pages 98–101 Assessment

1 a 835 cm, 8.4 m, 8405 mm b 80 m, 0.8 km, 80 500 cm c 20 300 cm, 2.03 km, 2300 m d 4.7 cm, 47 cm, 47.5 cm

2 a 37.68 cm b 97.34 cm c 84.78 cm d 59.66 cm e 73.79 cm f 52.75 cm

3 a 157 cm b 113.04 cm c 70.65 cm d 128.74 cm e 61.23 cm f 77.24 cm

4 a 8 cm b 5.5 cm c 11.5 cm d 2.5 cm

5

a

Length	Width	Area
10.5 cm	8 cm	84 cm^2
9 cm	6.5 cm	58.5 cm^2
5.25 cm	3 cm	15.75 cm^2

b

Length	Width	Area
11 cm	4.25 cm	46.75 cm^2
12.75 cm	7.5 cm	95.63 cm^2
15.5 cm	6.25 cm	96.875 cm^2

6 a 25.5 cm^2 b 63.25 cm^2 c 66 cm^2

7 a 40 cm^2 b 63 cm^2 c 406.25 cm^2

8 a 60 m^2 b 56 m^2 c 31.5 m^2

9 a 72 m^2 b 128 m^2 c 94.5 m^2

10 a 100 m^2 b 118.5 m^2

11 a 416 cm^3 b 330 cm^3 c 512 m^3 d 1050 m^3

12 a 220 cm^3 b 108 cm^3 c 432 m^3

13 a 350 cm^3 b 97 cm^3 c 2 m^3 d 6.5 m^3 e 8500 cm^3 f 3750 cm^3

14 a 30 mL b 2.4 L c 4000 L d 12 500 L e 915 mL f 7750 L

15 a 50° b 95° c 35° d 80° e 30° f 155°

16 a 100° b 145° c 145°

17 a 55° b 65° c 35° d 60° e 235° f 255°

18 a 325° b 080° c 225°

19 a S40°E b N75°E c S45°W d N20°W e S15°W f N40°E

20 a 1:200 000 b 1:500 c 1:50 000

21 a 750 m b 2 km c 625 m d 3.75 km

22 a 2 cm b 0.5 cm c 7 cm d 4.5 cm

23 a (–4, 2) b (–3, –1) c (3, 4) d (–2, 1) e (4, –2) f (–4, –5)

24 a square b circle c rectangle d triangle e hexagon f trapezium

Strand: Measurement

Page 102

1 a 4.13 kg b 2.405 kg c 4.885 kg d 5.06 t e 4.543 t f 5.806 kg g 2.44 kg h 287.288 kg i 196 g

2

Total weight	Number of items of the same weight	Weight of one item
2.184 kg	8	273 g
2.445 t	5	489 kg
4.578 kg	7	654 g
1.485 t	9	165 kg
5.08 kg	4	1.27 kg
43.197 kg	11	3.927 kg
4.776 kg	6	796 g

3 a 3.9 t b 1.5 t c 1.56 t d 1.575 t

4 a 24 kg b 57 kg c 40.6 kg d 22 kg

5 73 kg **6** 600 kg **7** 14 days

Page 103

1 21 to 5, 4:39 **2** 14 past 10, 10:14 **3** 8 past 5, 8:05

4 6 to 4, 3:54 **5** 27 past 12, 12:27 **6** 17 to 3, 2:43

7 22 past 6, 6:22 **8** 11 to 9, 8:49 **9** 26 to 1, 12:34

10 4 past 7, 7:04 **11** 28 to 11, 10:32 **12** 13 past 8, 8:13

13 19 to 2, 1:41 **14** 16 past 9, 9:16 **15** 24 to 10, 9:36

Page 104

1 a 42 days b 180 minutes c 84 months d 450 years
e 660 seconds f 50 years g 315 minutes h 42 months
i 270 seconds j 72 hours k 85 years l 465 minutes
m 84 days n 75 months o 228 hours p 45 years
q 56 days r 225 years s 140 minutes t 435 seconds
u 104 months v 32 years w 460 minutes x 575 years

2 a 6 hours b $2\frac{1}{2}$ hours c $4\frac{1}{2}$ hours d 10 hours
e 7 hours f $1\frac{1}{2}$ hours g $5\frac{1}{4}$ hours h $1\frac{3}{4}$ hours
i $2\frac{1}{4}$ hours j 9 hours

3 a 3 days b 10 days c 6 days d 5 days
e $1\frac{1}{2}$ days f $5\frac{1}{2}$ days g $8\frac{1}{2}$ days h $11\frac{1}{2}$ days
i $3\frac{1}{2}$ days j 7 days

4 a 4 years b 7 years c $1\frac{1}{2}$ years d $4\frac{1}{2}$ years
e $9\frac{1}{2}$ years f $5\frac{1}{2}$ years g $11\frac{1}{2}$ years h $1\frac{1}{4}$ year
i $3\frac{3}{4}$ years j $7\frac{1}{4}$ years

5 a 4 minutes b $3\frac{1}{2}$ minutes c $6\frac{1}{2}$ minutes d $1\frac{3}{4}$ minutes
e $4\frac{1}{4}$ minutes f $7\frac{1}{4}$ minutes g $9\frac{1}{4}$ minutes h $10\frac{1}{2}$ minutes
i $8\frac{3}{4}$ minutes j $12\frac{1}{2}$ minutes

6 a 6 centuries b 10 centuries c $4\frac{1}{2}$ centuries d $1\frac{1}{2}$ centuries
e $3\frac{1}{4}$ centuries f $9\frac{3}{4}$ centuries g 12 centuries h 50 centuries
i $6\frac{3}{4}$ centuries j $5\frac{1}{2}$ centuries

7 a 8 decades b 13 decades c $4\frac{1}{2}$ decades d $1\frac{1}{2}$ decades
e $6\frac{1}{5}$ decades f $9\frac{3}{5}$ decades g $3\frac{1}{10}$ decades h $17\frac{7}{10}$ decades
i $21\frac{1}{2}$ decades j $7\frac{1}{2}$ decades

Page 105

1 a 4:50 p.m. b 11:25 a.m. c 1:45 p.m. d 8:05 p.m.
e 9 a.m. f 2 p.m. g 7:36 a.m. h 5:27 p.m.
i 2:08 p.m. j 5:30 a.m. k 12:55 a.m. l 3:21 p.m.
m 7:34 p.m. n 2:54 a.m. o 10:41 p.m. p 11:59 p.m.
q 6:15 a.m. r 10:58 a.m. s 3:28 p.m. t 4:12 a.m.
u 12:07 p.m. v 6:19 p.m. w 9:52 p.m. x 12:26 a.m.

2 a 1615 b 1435 c 0923 d 0612 e 1700 f 1045
g 2308 h 1910 i 0748 j 0955 k 1205 l 0020
m 1318 n 1547 o 0526 p 2051 q 0439 r 0217
s 2104 t 1832

3 a 1518 b 10:40 p.m. c 0658 d 1350
e 1754 f 2230 g 1620 h 1850
i 10 p.m. j 1918 k 1425 l 11:15 p.m.
m 5:37 p.m. n 1356

4 a 1650, 1500, 2045, 2237, 1355 b 1123, 1004, 1355
c 1650 d 0735
e 0450, 0735, 0047 f 1650, 1500
g 0735, 1004
h 0047, 0450, 0735, 1004, 1123, 1355, 1500, 1650, 2045, 2237

Page 106

1 a 24 minutes b 53 minutes c 1 hour 5 minutes
d 8 minutes e 13 minutes f 42 minutes
g 50 minutes h 1 hour 1 minute

2 a 1 hour 35 minutes b 1 hour 52 minutes c 55 minutes
d 1 hour 1 minute e 18 minutes f 1 hour 14 minutes
g 25 minutes h 54 minutes

3 a 1235 b 1905 c 1125 d 2330 e 1600 f 0940
g 2010 h 1320 i 1750 j 0155 k 1045 l 2115

4 a $10\frac{1}{2}$ hours b $10\frac{1}{2}$ hours c $10\frac{1}{2}$ hours
d 9 hours 35 min e $9\frac{1}{2}$ hours f 7 hours 20 min

5 a 4 hours 37 min b 7 hours 35 min c 4 hours 44 min
d 9 hours 37 min e 7 hours 26 min f 7 hours 9 min
g 8 hours 51 min h 7 hours 21 min i 8 hours 35 min
j 6 hours 11 min k 5 hours 39 min l 9 hours 41 min
m 4 hours 12 min n 6 hours 18 min o 7 hours 22 min

Page 107

1 a 5 past 2 b 10 to 11 c 5 to 2
2 a 25 to 8 b 20 to 7 c $\frac{1}{4}$ to 9
3 a 5 past 9 b $\frac{1}{4}$ past 11 c $\frac{1}{4}$ past 9
4 a 5 to 8 b $\frac{1}{4}$ to 5 c 25 past 5
5 a $\frac{1}{2}$ past 8 b 20 past 12 c 25 to 9
6 a 5 to 3 b 5 past 1 c 25 past 3
7 a 0948 b 0643 c 0623
8 a 0046 b 0006 c 2231
9 a 1217 b 1532 c 1302

Page 108

1 1 minute 39 seconds
2 a 42 minutes b 8 hours 24 minutes
3 a 5 hours 24 minutes b 49 minutes c 36 minutes
4 1 hour 27 minutes
5 a $26\frac{1}{4}$ hours b 210 hours
6 a 2 hours 41 minutes b 10 hours 44 minutes c $11\frac{1}{2}$ hours
d 11 hours 53 minutes
7 a 1:15 p.m. b 2:57 p.m. c 2:40 p.m.
8 a K25.50 b K72.25 c K127.50 d K97.75
9 a 13 seconds b 8 minutes 26 seconds

Pages 109–110 Assessment

1 a 4.675 kg b 4.09 t c 6.262 kg d 441 g
2 a 10.56 kg b 14.715 kg
3 a 23 past 7, 7:23 b 9 to 5, 4:51 c 13 past 11, 11:13
4 a 4 hours b 66 months c $6\frac{1}{2}$ decades d 280 minutes
e 63 days f $4\frac{3}{4}$ centuries g 465 seconds h $6\frac{1}{2}$ years
i 264 hours
5 a 2:55 p.m. b 8:17 p.m. c 3:28 a.m. d 10:06 a.m.
e 6:42 p.m. f 10:34 p.m.
6 a 1946 b 1129 c 1312 d 0251
e 2153 f 1731 g 0905 h 1524
7 a 2:40 a.m., 1350, 1410, 2:17 p.m., 1436
b 5:50 a.m., 1640, 5:30 p.m., 1735, 1748
c 1010, 10:15 a.m., 10:50 a.m., 1055, 10:15 p.m.
d 8:10 a.m., 0820, 2035, 8:45 p.m., 2100
e 0615, 6:25 a.m., 1745, 6 p.m., 1815

8 a 1245 b 0730 c 1805 d 1455 e 0920 f 1140

9 a 11:12 a.m. b 1632 c 1007 d 2:22 p.m.
e 1737 f 12:17 p.m. g 3:27 p.m. h 2102

10 a 5 hours 20 minutes b 4 hours 38 minutes c 7 hours 12 minutes
d 7 hours 9 minutes e 5 hours 28 minutes f 7 hours 48 minutes

11 a 1839, 2054, 1959 b 4:55, 7:10, 6:15 c 0352, 0607, 0512
d 10:25, 12:40, 11:45 e 1223, 1438, 1343 f 4:00, 6:15, 5:20

Strand: Chance and Data

Page 111

1 a 43 b 44 c $59\frac{4}{5}$ d $55\frac{5}{6}$ e $66\frac{2}{3}$ f $67\frac{1}{2}$
g 8.3 h 7.7 i 9.1

2 a 10 b 10 c 10 d 15 e 15 f 15

3 a 36 b 73 c 28 d 64 e 42.5 f 89

4 a 16 b 6 c 7 d 8.5 e 3.4 f $8\frac{1}{2}$

5 a 78 b 44 c 26.1 d 26.7 e 6 f $8\frac{3}{4}$

Pages 112–114

1 a $\frac{1}{2}$ b $\frac{1}{3}$ c $\frac{1}{6}$ d $\frac{1}{2}$

2 a $\frac{1}{10}$ b $\frac{1}{2}$ c $\frac{1}{5}$ d $\frac{3}{10}$

3 a $\frac{1}{8}$ b $\frac{7}{8}$ c $\frac{1}{2}$ d $\frac{7}{8}$ e $\frac{1}{2}$ f $\frac{3}{8}$

4 a $\frac{3}{14}$ b $\frac{3}{7}$ c $\frac{4}{7}$

5 a $\frac{1}{4}$ b $\frac{3}{20}$ c $\frac{1}{2}$ d $\frac{1}{10}$

6 a $\frac{1}{2}$ b $\frac{1}{10}$ c $\frac{1}{5}$ d $\frac{4}{25}$ e $\frac{9}{50}$ f $\frac{7}{25}$

7 a $\frac{5}{12}$ b $\frac{5}{12}$ c $\frac{7}{12}$ d $\frac{7}{12}$

8 a $\frac{2}{5}$ b $\frac{1}{10}$ c $\frac{1}{2}$ d 0

9 a $\frac{1}{2}$ b $\frac{1}{3}$ c $\frac{1}{6}$ d $\frac{1}{6}$ e $\frac{1}{3}$ f $\frac{1}{3}$

10 a $\frac{1}{2}$ b $\frac{1}{2}$ c $\frac{1}{4}$ d $\frac{1}{4}$ e $\frac{1}{13}$ f $\frac{1}{13}$
g $\frac{1}{13}$ h $\frac{1}{26}$ i $\frac{3}{13}$ j $\frac{2}{13}$ k $\frac{3}{4}$ l $\frac{2}{13}$

11 a $\frac{1}{4}$ b $\frac{1}{4}$ c $\frac{3}{8}$ d $\frac{1}{8}$

12 a $\frac{9}{22}$ b $\frac{9}{22}$ c $\frac{5}{11}$ d $\frac{1}{11}$ e $\frac{7}{22}$ f $\frac{3}{22}$
g $\frac{5}{22}$ h $\frac{2}{11}$ i $\frac{7}{22}$ j $\frac{1}{11}$ k $\frac{1}{11}$ l $\frac{2}{11}$

13 a $\frac{5}{8}$ b $\frac{3}{8}$ c $\frac{3}{16}$ d $\frac{3}{8}$ e $\frac{1}{4}$ f $\frac{3}{16}$
g $\frac{3}{16}$ h $\frac{1}{8}$ i $\frac{3}{16}$ j $\frac{5}{16}$ k $\frac{1}{4}$ l $\frac{5}{16}$

Page 115

1 a 13.501 b 56.16 c 47.476 d 58.695 e 207.48 f 194.91
g 79.202 h 324.988 i 48.846 j 312.625 k 219.704 l 216.068
m 156.915 n 62.964 o 1.416

2 a smaller b larger c larger d smaller e larger f smaller
g larger h smaller i larger j smaller k smaller l larger
m larger n smaller o larger p larger q larger r larger
s larger t larger

3 a 45×0.99 b $26 \div 0.9$ c $32 \div 1.2$ d 4×0.56
e 15×0.98 f $50 \div 1.6$ g 25×0.9 h $11 \div 1.1$
i 18×0.03 j 12×1.75 k $24 \div 1.25$ l $10 \div 0.88$
m 16×0.04 n $30 \div 1.8$ o $45 \div 1.4$ p 36×0.75
q $20 \div 1.25$ r 9×1.01 s $28 \div 0.5$ t 5×0.07
u $40 \div 0.87$

4 a 2.5 b 12.25 c 11.52 d 35.7 e 58 f 8.2
g 19.5 h 5.6 i 21.8 j 30.9 k 25 l 5.04
m 27.2 n 14.48 o 14.64

Page 116

1 Estimates will vary.
a K46.25 b K31 c K24.75 d K23.40 e K22.80 f K36.45
g K38.80 h K42.90 i K39.15 j K89.10 k K86 l K127.60

2 Estimates will vary.
a K25.50 b K16.80 c K34 d K11.55 e K4.80 f K14.75
g K17.25 h K24.85 i K7.50 j K11.50 k K17.15 l K24.75
m K24.15 n K51.75 o K49.50 p K7 q K16.80 r K6
s K3

Page 117 Assessment

1 a $21\frac{3}{4}$ b $22\frac{1}{2}$ c 40

2 a 3.8 b 94 c $10\frac{1}{2}$ and $9\frac{1}{4}$

3 a $\frac{1}{2}$ b $\frac{3}{4}$ c $\frac{1}{4}$ d $\frac{1}{8}$

4 a $\frac{7}{10}$ b $\frac{9}{50}$ c $\frac{7}{50}$ d $\frac{17}{25}$ e $\frac{6}{25}$ f $\frac{1}{5}$
g $\frac{1}{10}$ h $\frac{1}{10}$

5 a 27×1.3 b 16×0.98 c $70 \div 2.08$ d $36 \div 0.5$
e 52×0.75 f 23×0.85

6 a smaller b larger c smaller d larger
e larger f smaller g smaller h larger

7 The following answers are estimates to the nearest whole kina.
a K45 b K54 c K144 d K40 e K54 f K165

Strand: Patterns and Algebra

Page 118

1 12, 22, 15, 17, 30, 26

2 18, 13, 31, 49, 37, 25

3 44, 32, 60, 88, 72, 144

4 10, 7, 16, 9, 5, 15

5 12, 20, 9, 22, 26, 7

6 60, 45, 75, 30, 125, 100

7 33, 12, 66, 27, 138, 99

8 58, 170, 82, 106, 210, 266

9 5, 9, 18, 14, 3, 16

10 31, 39, 21, 57, 111, 81

11 60, 88, 44, 160, 124, 64

12 80, 230, 160, 490, 300, 410

13 106, 70, 406, 150, 87, 127

14 13, 11, 17, 16, 14, 22

Page 119

1 Add 4 to each IN number.

IN	15	11	23	32	37	21
OUT	19	15	27	36	41	25

2 Subtract 8 from each IN number.

IN	27	35	17	21	41	60
OUT	19	27	9	13	33	52

3 Multiply each IN number by 5.

IN	11	8	15	9	18	12
OUT	55	40	75	45	90	60

4 Divide each IN number by 7.

IN	56	14	42	21	77	63
OUT	8	2	6	3	11	9

5 Double each IN number and then add 8.

IN	10	25	16	26	20	14
OUT	28	58	40	60	48	36

6 Halve each IN number and then subtract 5.

IN	32	46	22	38	62	44
OUT	11	18	6	14	26	17

7 Add 7 to each IN number and then double.

IN	8	20	17	13	25	22
OUT	30	54	48	40	64	58

8 Multiply each IN number by 3 and then subtract 1.

IN	11	14	7	10	22	30
OUT	32	41	20	29	65	89

9 Multiply each IN number by 4 and then add 6.

IN	12	25	18	14	9	8
OUT	54	106	78	62	42	38

10 Halve each IN number and then add 3.

IN	30	76	16	24	42	54
OUT	18	41	11	15	24	30

11 Subtract 2 from each IN number and then divide by 4.

IN	30	18	46	82	38	62
OUT	7	4	11	20	9	15

12 Add 5 to each IN number and then multiply by 4.

IN	20	30	15	12	9	23
OUT	100	140	80	68	56	112

13 Divide each IN number by 5 and then add 1.

IN	25	50	60	75	45	35
OUT	6	11	13	16	10	8

14 Double each IN number and then subtract 4.

IN	14	30	21	44	17	16
OUT	24	56	38	84	30	28

15 Subtract 10 from each IN number and then multiply by 3.

IN	40	18	23	22	31	35
OUT	90	24	39	36	63	75

16 Multiply each IN number by itself and then add 2.

IN	10	4	7	12	20	9
OUT	102	18	51	146	402	83

Page 120

1 Divide by 10.
2 Multiply by 8.
3 Subtract 12.
4 Add 25.
5 Multiply by 3.
6 Subtract 7.
7 Divide by 5 and then add 1.
8 Multiply by 10 and then subtract 2.
9 Double and add 1.
10 Divide by 2 and then subtract 1.
11 Subtract 1 and then multiply by 3 or multiply by 3 and then subtract 3.
12 Add 2 and then halve or halve and then add 1.
13 Square the number.
14 Halve and then add 5.
15 Double and then subtract 1.
16 Multiply by 5 and then add 1.
17 Add 11.
18 Multiply by 15.

Page 121

1 Multiply by 8; OUT numbers are 72 and 320
2 Subtract 20; OUT numbers are 33 and 47
3 Add 50; OUT numbers are 84, 68 and 76
4 Multiply by 9; OUT numbers are 63, 108 and 81
5 Subtract 9; OUT numbers are 83, 17 and 49
6 Divide by 4; OUT numbers are 3, 7 and 18
7 Add 21; OUT numbers are 54, 50 and 66
8 Halve; OUT numbers are 7 and 35; IN number is 44
9 Multiply by 10 and then subtract 1; OUT numbers are 29, 209 and 109
10 Multiply by 10 and then add 1; OUT numbers are 41, 91 and 231
11 Double; OUT numbers are 38 and 132; IN numbers are 27 and 31
12 Double and then add 5; OUT numbers are 37, 15 and 63
13 Halve and then subtract 1; OUT numbers are 20, 32 and 14
14 Double and then subtract 10; OUT numbers are 28, 62 and 46
15 Halve and then add 5; OUT numbers are 40, 21 and 50
16 Square; OUT numbers are 900 and 256; IN number is 11
17 Find the square root; OUT numbers are 2 and 10; IN number is 49
18 Multiply by 10 and add 3; OUT numbers are 23, 393 and 613

Page 122

1 **a** $y = d - 10$ **b** $w = g + 12$ **c** $m = s + 5$ **d** $a = b - 8$
e $c = 20 - h$ **f** $f = 100 + p$

2 **a** $a = 7b$ **b** $h = 11k$ **c** $f = 4g$ **d** $w = 5y$
e $d = 10c$ **f** $n = 8p$

3 **a** $m = 4a - 2$ **b** $x = 5(y + 7)$ **c** $b = 2c + 12$ **d** $d = f \div 4 - 1$
e $p = (q - 3) \div 5$ **f** $f = 6g - 10$ **g** $s = 9(r \div 2)$ **h** $h = 10(j - 9)$

4 **a** $y = x + 9$ **b** $y = x - 15$ **c** $y = 5x + 2$ **d** $y = (x + 2) \div 5$
e $y = 3(x - 8)$ **f** $y = x \div 5 - 2$ **g** $y = 10x - 1$ **h** $y = (x - 4) \div 2$

5 **a** To find *y*, subtract 8 from *w*. **b** To find *x*, add 4 to *z*.
c To find *s*, multiply *y* by 9. **d** To find *h*, multiply *f* by 5.
e To find *g*, multiply *h* by 2 and then subtract 7.
f To find *a*, add 8 to *b* and then multiply by 3.
g To find *d*, add 5 to *c* and then divide by 3.
h To find *k*, subtract 3 from *m* and then multiply by 7.
i To find *p*, divide a by 3 and then add 2.
j To find *r*, subtract 2 from *s* and then multiply by 6.
k To find *b*, subtract *w* multiplied by 3 from 20.
l To find *v*, multiply *x* by 8 and then add 7.

Page 123

1 $y = 15, 33, 7, 68, 46, 29$
2 $y = 32, 64, 44, 24, 80, 400$
3 $y = 20, 29, 36, 5, 9, 23$
4 $y = 54, 31, 45, 79, 40, 101$
5 $y = 25, 53, 39, 69, 63, 97$
6 $y = 39, 115, 27, 55, 91, 79$
7 $y = 100, 25, 64, 144, 625, 81$
8 $y = 36, 86, 32, 48, 70, 60$
9 $y = 50, 15, 70, 100, 30, 125$
10 $y = 36, 30, 8, 22, 32, 26$
11 $y = 29, 19, 23, 16, 10, 22$
12 $y = 36, 56, 106, 71, 161, 116$
13 $y = 30, 21, 3, 75, 27, 39$
14 $y = 8, 125, 1000, 27, 512, 216$
15 $y = 17, 19, 14, 11, 22, 38$
16 $y = 115, 43, 10, 75, 138, 19$

Page 124

1 **a** 13 **b** 30 **c** 1 **d** 9 **e** 24 **f** 11
g 4 **h** 25 **i** 21 **j** 32 **k** 28 **l** 39

2 **a** 42 **b** 55 **c** 19 **d** 11 **e** 73 **f** 44
g 24 **h** 33 **i** 4 **j** 55 **k** 49 **l** 192

3 **a** 10 **b** 2 **c** 14 **d** 24 **e** 48 **f** 36
g 14 **h** 4 **i** 50 **j** 20 **k** 32 **l** 33

4 **a** 63 **b** 5 **c** 43 **d** 126 **e** 2 **f** 77
g 27 **h** 40 **i** 31 **j** 35 **k** 6 **l** 64

5 **a** 10 **b** 4 **c** 11 **d** 13 **e** 30 **f** 32
g 9 **h** 30 **i** 70 **j** 20 **k** 8 **l** 4

6 a 12 b 4 c 62 d 80 e 39 f 240
g 32 h 66 i 66 j 16 k 76 l 24

7 a 66 b 102 c 21 d 111

8 a 19 b 7 c 25 d 45

9 a 34 b 112 c 25 d 13

Page 125 Assessment

1 a 17, 9, 31, 52, 33, 20 b 60, 40, 124, 72, 200, 108
c OUT numbers: 6, 12, 9; IN numbers: 24, 8, 60
d OUT numbers: 41, 23, 67: IN numbers: 13, 22, 18

2 a Multiply by 6. b Add 19.

3 a $a = b + 20$ b $x = 7y$ c $g = 18 - h$ d $w = 4z$
e $c = 8d + 2$ f $k = (m - 5) \div 3$ g $e = 9(f + 2)$ h $p = 5(q \div 4)$

4 Rules should be similar to these ones and have the same meaning.
a To get *z*, multiply *w* by 3 and then subtract 1.
b To get *a*, multiply *b* by 11.
c To get *g*, add 5 to *h* and then multiply by 2.
d To get *s*, multiply *t* by 2 and then subtract the answer from 14.

5 a 29 b 15 c 36 d 26

6 a 81 b 39 c 129 d 105 e 210 f 171

7 a 28 b 12 c 48 d 32 e 60 f 200

Oxford University Press is a department of the University of Oxford. It furthers the University's objective of excellence in research, scholarship, and education by publishing worldwide. Oxford is a registered trademark of Oxford University Press in the UK and in certain other countries.

Published in Australia by
Oxford University Press
253 Normanby Road, South Melbourne, Victoria 3205, Australia

First published 2011
Reprinted 2020(D)

ISBN 978 0 19 557600 9

Edited by Emma Short
Typeset by Palmer Higgs
Printed and bound in Australia by Ligare Book Printers Pty Ltd